Innovative Solutions and Challenges in Civil Engineering (ISCCE 2020)

IOP Conference Series: Materials Science and Engineering
Volume 1026

Online
20 - 21 August 2020

ISBN: 978-1-7138-3326-0
ISSN: 1757-8981

Printed from e-media with permission by:

Curran Associates, Inc.
57 Morehouse Lane
Red Hook, NY 12571

Some format issues inherent in the e-media version may also appear in this print version.

Institute of Physics
Dirac House, Temple Back
Bristol BS1 6BE UK

Phone: 44 1 17 929 7481
Fax: 44 1 17 920 0979

techtracking@iop.org

Additional copies of this publication are available from:

Curran Associates, Inc.
57 Morehouse Lane
Red Hook, NY 12571 USA
Phone: 845-758-0400
Fax: 845-758-2633
Email: curran@proceedings.com
Web: www.proceedings.com

TABLE OF CONTENTS

WATER RESOURCES MANAGEMENT

STRUCTURAL ENGINEERING

GEOTECHNICAL ENGINEERING

CONSTRUCTION ENGINEERING AND MANAGEMENT

IOP Conf. Series: Materials Science and Engineering 1026 (2021) 011001 doi:10.1088/1757-899X/1026/1/011001

Preface

Department of Civil Engineering, Saveetha Engineering College, Chennai in association with IEI Student chapter has organized a Virtual International Conference on INNOVATIVE SOLUTIONS AND CHALLENGES IN CIVIL ENGINEERING on 21 st and 22 nd August 2020. The two day international conference has provided opportunity to enrich and share the knowledge and novel ideas for sustainable development of the society. Renowned speakers from various countries such as Australia, Africa, Kuwait, USA, shared their knowledge in their area of expertise. The conference invited professionals, scientists, academicians, research scholars, industrialists and students to contribute their original research work in the areas but not limited to the following themes.

Theme1. Structural and Construction Engineering

Theme2. Geotechnical and Geological Sciences.

Theme3. Environmental and Water Resources Management

Theme4. Transportation Engineering

Theme5. 3D Printing

Theme6. Artificial Intelligence & Internet of Things (IOT)

Theme7. Public health and Awareness

The Key Note Speakers for our virtual conference are the following

Dr. Muhammed Hadi

Associate Professor, University of Wollongong, New South Wales, Australia.

Dr. Mbereyaho Leopold

College of Science and Technology, University of Rwanda, Africa

Dr.S.Sreedeep

Professor, Geotechnical Engineering, IIT Guwahati

Dr. Jafarali Parol

Energy and Building Research Centre, Kuwait Institute for Scientific Research, Kuwait

Dr. Ganesh Gopalakrishnan

Institution of Oceanography, UC San Diego, California, USA.

To ensure the quality during the conduct of virtual conference, we had taken suggestions from expert committee comprising of the following members

Dr.Kartini Kamaruddin, Associate Professor, Universiti Teknologi MARA, Malaysia.

Dr.G.Dodagoudar, Professor, IIT Madras.

Dr.B.V.Mudgal, Professor & Director, Centre for Water Resources, Anna University, Chennai.

Dr.S.R.Sathish Kumar, Professor, IIT Madras.

Dr.Y.R.M.Rao, Principal, Dr.Paul's Engineering College, Villupuram Dt. Tamil Nadu.

Dr.S.Thirugnanasambandam, Professor, Annamalai University.

Dr.Sreeja Pekkat, Associate Professor, IIT Guwahati.

Dr.K.Krishnamurthy, Professor, NIT Calicut.

Dr.Rakesh J.Pillai, Assistant Professor, IIT Palakkad.

Dr.S.Saravanan, Assistant Professor, NIT Trichy.

We received 40 conference papers and the papers were reviewed and scrutinized for the technical relevance and content by the following editorial team. The papers are checked for plagiarism using the software "iThenticate".

1. Dr. Jailakshmi Menon, Associate Professor and Head, Saveetha Engineering College, Chennai
2. Dr. Subathra Devi V, Professor, Saveetha Engineering College, Chennai
3. Dr. Kanagarathinam L, Associate Professor, Saveetha Engineering College, Chennai

The editorial team selected 17 papers for publication.

Peer review declaration

All papers published in this volume of *IOP Conference Series: Materials Science and Engineering* have been peer reviewed through processes administered by the Editors. Reviews were conducted by expert referees to the professional and scientific standards expected of a proceedings journal published by IOP Publishing.

- **Type of peer review:**
  Single- blind review process.
  Reviewers know the identity of authors but authors do not know the identity of reviewers.
- **Conference submission management system:** Easy chair
- **Number of submissions received:** 34
- **Number of submissions sent for review:** 22
- **Number of submissions accepted:** 17
- **Acceptance Rate (Number of Submissions Accepted / Number of Submissions Received X 100):** 50%
- **Average number of reviews per paper:** 01
- **Total number of reviewers involved:** 4 (Per discipline)
- **Any additional info on review process:** All the accepted papers are checked for plagiarism using the software **"iThenticate ".**
- **Contact person for queries:** Dr.Jailakshmi menon, HoD & Convenor, Dr.V.Subathra Devi, Conference secretary.

IOP Conf. Series: Materials Science and Engineering 1026 (2021) 012001 doi:10.1088/1757-899X/1026/1/012001

Impact of Fragile Water Management Strategies and Mitigation-A Case Study of Pune City, India

Dr. J S Sudarsan[1], Kakuru Jyothi Priyanka Reddy[2], Haseeb.A.H. Biyabani[3] Purnima Kumari[4] and Swati Sinha[5]

[1,2,3,4,5]ANational Institute of Construction Management and Research, Pune, Maharashtra, India.

sudarsanjss@gmail.com[1], kakurujyothi140997@gmail.com[2], haseebbiyabani5@gmail.com[3], 4purnima25@gmail.com[4], swatisinha.hi@gmail.com[5]

Abstract. The aim of the study is to know about the impact of water management system on the city of Pune due to the fluctuating monsoon. To achieve this, research study was carried out in exploring the factors which are contributing for this variation in amount of precipitation. For instance, improper urbanization and industrialization in the states such as that of Maharashtra is happening in a haphazard manner owing to the uncontrolled burst in population in the cities and town. To ascertain the impact caused due to heavy rainfall this study was initiated. Data of population of Pune city was collected as it is our study area for research. Based on the analysis it was clear that there is a drastic increment in last three decades. A survey was conducted and responses from people across the city were collected regarding the management of rain water as well as the issues faced during peak rainfall and it was analyzed to evaluate the impact. For analysis and identifying the regions that were most affected, SPSS (**S**tatistical **P**ackage for the **S**ocial **S**ciences) tool has been used. The survey responses collected were analyzed using SPSS and various parameters have been compared to identify the relevant ones. Based on the analysis of the data obtained from the survey it is concluded that in many regions proper rain water management strategies are not adopted and also rain water harvesting technique is not installed. This results in various problems like water logging, traffic congestion and improper drainage. The output obtained is conclusive on identifying the parameters that have crucial impact on the city during monsoons and if corrected, can bring a change in the rain water management strategies adopted and make Pune a better city to live in.

1. Introduction

During the last few years, increased attention is given on the issue of sustainability of urban water systems from ecological, social and economic point of views. In metro cities of India like Pune, it is difficult challenge to provide a sustainable underground storm water management system due to space constraint. The rain water should be managed properly and diverted effectively toward catchment or basins to reduce the waterlogging problems in peak rainy seasons. Due to improper urbanization and industrialization the development in the states such as that of Maharashtra is happening in a haphazard manner owing to the uncontrolled burst in population in the metro cities and town and even due to growing migration of population from the villages [1]. All these factors are resulting in the environmental degradation which affects the season cycle of our planet. Therefore, it leads to adverse effects such as rise in temperature, fluctuation in monsoon, heavy rainfall etc. This study focus on how the variation in North-East monsoons is affecting the normal people living in the city, what problems they are facing during rainy season and suggesting remedial measures to mitigate it. There are many techniques and measures that has been implemented by the government authorities to manage rain water during peak rainy seasons but these measures are not sufficient or may be not efficient as predicted [2]. Therefore, it is now the responsibility of each and every individual living in the city to take some serious action against this issue for controlling the factors which affect the environment and also to conserve water for building a sustainable environment for living [3].

As urbanization is increasing in most of the cities in India rapidly including Pune city, paved areas in the city is also increasing which leads to less infiltration of the rain water. During past few years amount of rainfall is varying significantly as given in Figure 1. Therefore, increasing the load on the storm water drainage network of the city. This improper draining of runoff water results in accumulation of water on roads and different parts of the city [4].

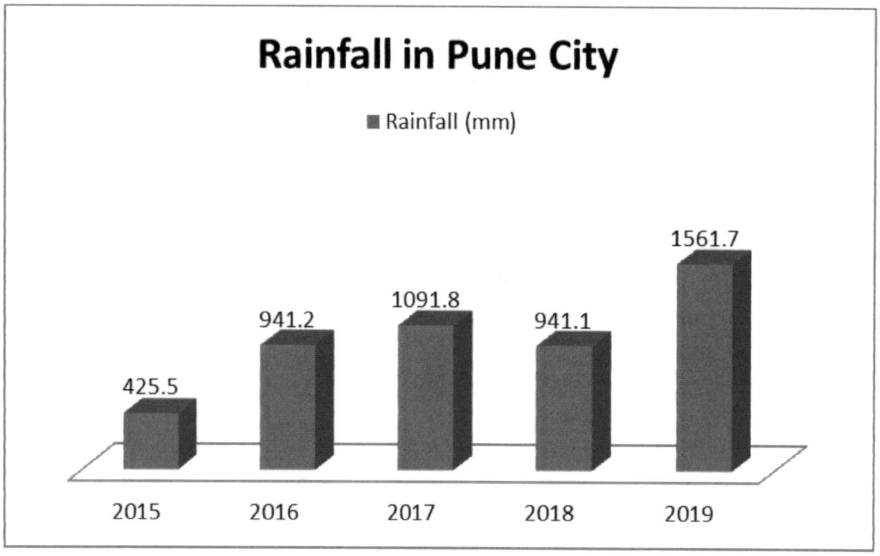

Figure 1. Amount of rainfall in Pune City

According to an article in Hindustan Time, Pune Municipal Corporation faced huge difficulties in 2018 to manage the rain water as they were not prepared for the heavy pre- monsoon rains. Roads were water logged in many areas of the city due to which residents faced lot of problems. People faced many issues like traffic congestion, choking of drains, areas were flooded and water entered into the houses [5]. In Pune (Maharashtra) population has increased three-folds in last three decades resulting in increased paved areas which leads to increased runoff. The city, on the other hand receives heavy rainfall causing huge runoffs because of lack of pervious surfaces and insufficient drainage system resulting in water logging [6]. The fluctuating amount of rainfall in past few years in Pune put additional pressure on the storm water management system as represented in figure 1. Also due to significant rise in population there is a substantial rise in siphoning of underground water which brought about diminishing the level of ground water table [7]. Pune authorities are continuously struggling to control this problem of mismanagement of rain water [8]. So, the main purpose of this study is to study impact of monsoon in Pune city due to fragile management of authorities and suggesting suitable remedy for the same.

2. METHODOLOGY

The methodology adopted in this work is determining various factors responsible for various environmental changes such as monsoon fluctuation and its impact on the people living in the city. Thus, the objective of current study is to analyze the variation in amount of rainfall happening in past few years and due to this what all problems are arising. Also suggesting some of the remedial measures as stated in Figure 2.

1. Since last few years' lot of variations can be observed in the pattern of rainfall in various regions of the country. There are several reasons for this change like pollution, global warming and other human activities that disturbs the environmental as well as geological conditions. Due to urbanization more and more industries are established in past few years, attracting more people for immigration. This swiftly expanding urbanization in various parts of India has led to generation of enormous quantity of air and water pollution causing global warming. All these factors causing negative environmental impacts like change in monsoon pattern, deteriorating the quality of fresh water, increasing temperature, varied rate of precipitation, and most vital is declining the ground water level. Most of the states in India are growing rapidly like Maharashtra, due to which paved areas in the state is also increasing leading to less infiltration of the rain water. Growing construction activities increases the pollution and also allows the rain water to flow directly into the drainage or streams rather than infiltrating into the soil which in turn is reducing the level of ground water table.

   The Areas which are highly affected during monsoon. In case of primary source, we have conducted a socio-economic survey in different areas of Pune city. In this survey we have prepared a questionnaire and approached to the people living in different areas of Pune city like Aundh, Baner, Pashan, Shivajinagar, etc. Goal of conducting the survey is to know that what type of problems occur due to heavy rainfall and up to which extent people living in an area is suffering.
2. After collecting sufficient number of reliable responses, we have segregated the data based upon the area in which an individual live. We have used SPSS (Statistical Package for the Social Sciences) tool for analysing the data. We conducted the Chi-Square test to obtain the relationship between the selected parameters for each region which helped us to reach to our desired results.

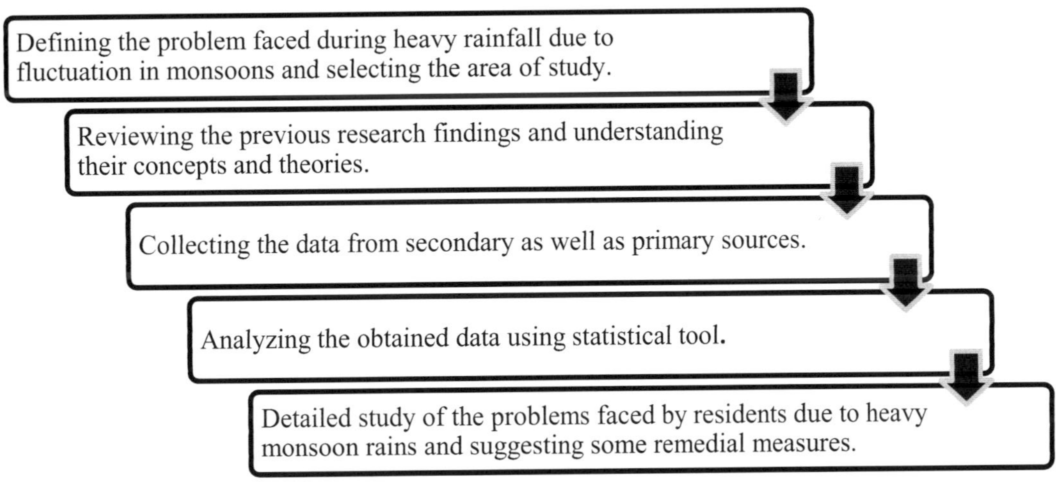

Figure 2. Research Methodology

3. Based upon the obtained results by analysing the primary data as well as taking the reference from the secondary data we have done a detailed research that how the tremendous rainfall has affected the residents of Pune city. How the people of the city are suffering due to different problems like traffic congestion, water logging and floods. They are still fighting with these issues and trying to find an efficient solution for it. We studied about what measure have been taken till now to mitigate these problems during rainy season and we have also suggested some remedial measure as well that can help to improve the situation.

3.STUDY AREA DESCRIPTION

Pune district occupies a strategic position within western Maharashtra. The district headquarters in Pune city. Pune is the second largest city in the Indian state of Maharashtra, after Mumbai. It is the ninth most populous city in the country with an estimated population of 3.13 million. Along with its extended city limits Pimpri-Chinchwad and the three cantonment towns of Pune, Khadki and Dehu Road, Pune forms the urban core of the eponymous Pune Metropolitan Region (PMR).

Pune city's total area is 15.642 sq. km. By road Pune is 1,173 km (729 mi) south of Delhi, 734 km (456 mi) north of Bangalore, and 149 km (93 mi) south-east of Mumbai. Pune lies on the western margin of the Deccan plateau, at an altitude of 560 m (1,840 ft.) above sea level.

Pune has a hot semi-arid climate bordering with tropical wet and dry with average temperatures ranging between 20 and 28 °C (68 and 82 °F). Pune experiences three seasons: summer, monsoon, and winter. Typical summer months are from mid-March to mid-June, with maximum temperatures sometimes reaching 42 °C (108 °F). The warmest month in Pune is May. The city often has heavy dusty winds in May, with humidity remaining high. Even during the hottest months, the nights are usually cool due to Pune's high altitude. The highest temperature recorded was 43.3 °C (109.9 °F) on 30 April 1897. The monsoon lasts from June to October, with moderate rainfall and temperatures ranging from 22 to 28 °C (72 to 82 °F). Most of the 722 mm (28.43 in) of annual rainfall in the city falls between June and September, and July is the wettest month of the year.

3.1 Demographic

The city has a population of 3,124,458; while 5,057,709 people reside in the Pune Urban Agglomeration as of the 2011 census. The latter was 4,485,000 in 2005. According to the Pune Municipal Corporation (PMC), 40% of the population lived in slums in 2001 The number of people migrating to Pune rose from 43,900 in 2001 to 88,200 in 2005. The sharp increase in population during the decade 1991-2001 led to the absorption of 38 fringe villages in and around Pune city.

4. RESULTS AND DISCUSSION

4.1 Primary Data Collected

Primary data has been collected by means of a socio-economic survey that was circulated in both software and hardware form. Printed forms were circulated and filled by people from various locations. Google form link was shared via email to people to collect responses. Social media such as WhatsApp was used to circulate the google form to various people. The target people for this survey were strictly people living in the city of Pune under PMC from various locations. A total of 60 responses were obtained and used for further analysis as represented in the figure 3,4 and table 1,2.

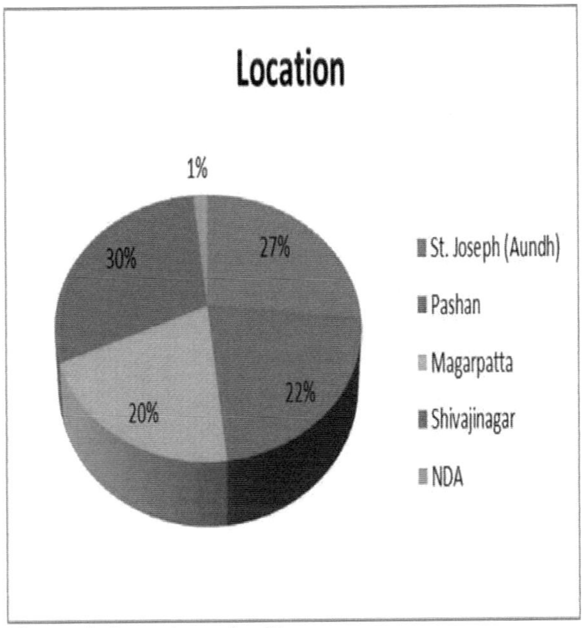

Figure 3. Location wise Responses

Location	No.of responses
St. Joseph (Aundh)	16
Pashan	13
Magarpatta	12
Shivajinagar	18
NDA	1
Total	60

Table 1. Location wise Responses

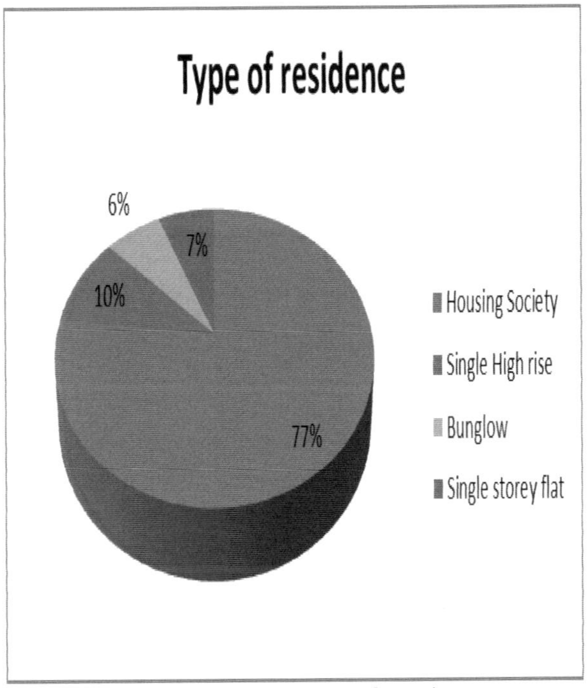

Figure 4. Responses as per type of residence
residence

Type of residence	No. of responses
Housing Society	46
Single High rise	6
Bunglow	4
Single storey flat	4

Table 2. Responses as per type of

Based on the data collected only two locations (Aundh and Pashan) were selected for further discussion and analysis as these locations are highly affected.

4.2 Observations:

- It has been a dwelling area in Pune from a long time and is still developing. Around 30% of the people are living here for more than 10 years and another 30% of the people are new dwellers in the area.
- It is a well-developed area with demand for living here as 75% of the homes here are in housing societies. This reflects a very high population density in this area.
- However, only around 30% of the societies have rain water harvesting system installed in their Apartments.
- Water logging is not a usual occurrence in this area, however, close to 60% people experienced water logging this monsoon. 75% of the population faced various problems due to several factors.
- Approximately 20% of the area gets cleared of water logging within 24 hours whereas 40% of the area faces water logging for more than several hours.

4.3 SPSS Analysis:

Test1: To check the relation between rain water harvesting facility available or installed and experiencing water logging in the area.

Null hypothesis -There is no relation between rain water harvesting facility available or installed and experiencing water logging in that area.

Alternate hypothesis -There is relation between rain water harvesting facility available or installed and experiencing water logging in that area.

	Valid		Missing		Total	
	N	Percent	N	Percent	N	Percent
Is rain water harvesting facility installed/available in your residence? * Did you experience water-logging in your area?	16	100.0 %	0	0.0 %	16	100.0 %

Table 3. Case Processing Summary

			Did you experience water-logging in your area?		
			No	Yes	Total
Is rain water harvesting facility installed/available in your residence?	No	Count	3	8	11
		Expected Count	4.8	6.2	11.0
	Yes	Count	4	1	5
		Expected Count	2.2	2.8	5.0
Total		Count	7	9	16
		Expected Count	7.0	9.0	16.0

Table 4. Is rain water harvesting facility installed/available in your residence? * Did you experience water-logging in your area? Cross tabulation.

	Value	df	Asymptotic Significance (2-sided)	Exact Sig. (2-sided)	Exact Sig. (1-sided)
Pearson Chi-Square	3.883[a]	1	.049		
Continuity Correction[b]	2.036	1	.154		
Likelihood Ratio	4.035	1	.045		
Fisher's Exact Test				.106	.077
N of Valid Cases	16				

Table 5. Chi-Square Tests

4.4 Inferences:
Since the Pearson Chi-Square value for Test 1 (=0.049) is less than 0.05, hence null hypothesis is rejected and the alternate hypothesis is accepted as per above table 5.
Therefore, it is inferred that there is relation between rain water harvesting facility available or installed and experiencing water logging in an area

Test 2: To check the relation between rain water harvesting facility available or installed and time required for the water logging to subside.

Null hypothesis – There is no relation between rain water harvesting facility available or installed and time required for the water logging to subside.

Alternate hypothesis – There is relation between rain water harvestingfacility available or installed and time required for the water logging to subside.

	Cases					
	Valid		Missing		Total	
	N	Percent	N	Percent	N	Percent
Is rain water harvesting facility installed/available in your residence? * How long does it take for the water-logging to subside?	16	100.0%	0	0.0%	16	100.0%

Table 6. Case Processing Summary

			How long does it take for the water-logging to subside?				
			24 - 48 hours	48 - 72 hours	Less than 24 hours	More than 72 hours	Total
Is rain water harvesting facility installed/available in your residence?	No	Count	2	4	0	5	11
		Expected Count	2.1	2.8	2.1	4.1	11.0
	Yes	Count	1	0	3	1	5
		Expected Count	.9	1.3	.9	1.9	5.0
Total		Count	3	4	3	6	16
		Expected Count	3.0	4.0	3.0	6.0	16.0

Table 7. Is rain water harvesting facility installed/available in your residence? * How long does it take for the water-logging to subside? Cross tabulation

	Value	df	Asymptotic Significance (2-sided)
Pearson Chi-Square	9.018[a]	3	.029
Likelihood Ratio	10.649	3	.014
N of Valid Cases	16		

Table 8. Chi-Square Tests

Inferences:

- Since the Pearson Chi-Square value for Test 1 (=0.029) is less than 0.05, hence null hypothesis is rejected and the alternate hypothesis is accepted as per above TABLE 4.15
- Therefore, it is evident that relation between rain water harvesting facility available or installed and time required for the water logging to subside.

Pashan

Observations:

- It is a relatively new dwelling location for the people of Pune with around 75% people living here for less than 5 years duration in this location.
- It is a rapidly developing area with around 75% of the residences being housing societies-based data collected.
- Around 50% of the residences do not have rain water harvesting facility installed Derived from the data collected.
- This location does not usually experience water logging but 55% of the people experienced water logging this monsoon and faced various problems.
- Around 50% of the area is relieved of water logging within 24 hours whereas the other half requires more than 72 hours. This indicates absence of proper drainage system in the other half

and can be relieved of this problem by providing a proper drainage system.

SPSS Analysis:

Test1: To check the relation between rain water harvesting facility available or installed and experiencing water logging in the area.
Null hypothesis – There is no relation between rain water harvesting facility available or installed and experiencing water logging in that area.
Alternate hypothesis – There is relation between rain water harvesting facility available or installed and experiencing water logging in that area.

	Cases					
	Valid		Missing		Total	
	N	Percent	N	Percent	N	Percent
Is rain water harvesting facility installed/available in your residence? * Did you experience water-logging in your area?	13	100.0%	0	0.0%	13	100.0%

Table 9. Case Processing Summary

			Did you experience water-logging in your area?		
			No	Yes	Total
Is rain water harvesting facility installed/available in your residence?	No	Count	1	5	6
		Expected Count	2.8	3.2	6.0
	Yes	Count	5	2	7
		Expected Count	3.2	3.8	7.0
Total		Count	6	7	13
		Expected Count	6.0	7.0	13.0

Table 10. Is rain water harvesting facility installed/available in your residence? *Did you experience water-logging in your area? Cross tabulation

	Value	df	Asymptotic Significance (2-sided)	Exact Sig. (2-sided)	Exact Sig. (1-sided)
Pearson Chi-Square	3.899[a]	1	.048		
Continuity Correction[b]	2.006	1	.157		
Likelihood Ratio	4.162	1	.041		
Fisher's Exact Test				.103	.078
N of Valid Cases	13				

Table 11. Chi-Square Tests

Inferences:

- Since the Pearson Chi-Square value for Test 1 (=0.048) is less than 0.05, hence null hypothesis is rejected and the alternate hypothesis is accepted based on TABLE 4.25
- Therefore, it is clear that relation between rain water harvesting facility available or installed and experiencing water logging in an area.

Test2: To check the relation between rain water harvesting facility available or installed and time required for the water logging to subside.

Null hypothesis – There is no relation between rain water harvesting facility available or installed and time required for the water logging to subside.

Alternate hypothesis – There is relation between rain water harvesting facility available or installed and time required for the water logging to subside.

	Cases					
	Valid		Missing		Total	
	N	Percent	N	Percent	N	Percent
Is rain water harvesting facility installed/available in your residence? * How long does it take for the water-logging to subside?	13	100.0%	0	0.0%	13	100.0%

Table 12. Case Processing Summary

			How long does it take for the water-logging to subside?		
			Less than 24 hours	More than 72 hours	Total
Is rain water harvesting facility installed/available in your residence?	No	Count	1	5	6
		Expected Count	2.8	3.2	6.0
	Yes	Count	5	2	7
		Expected Count	3.2	3.8	7.0
Total		Count	6	7	13
		Expected Count	6.0	7.0	13.0

Table 13. Is rain water harvesting facility installed/available in your residence? * How long does it take for the water-logging to subside?

	Value	df	Asymptotic Significance (2-sided)	Exact Sig. (2-sided)	Exact Sig. (1-sided)
Pearson Chi-Square	3.899[a]	1	.048		
Continuity Correction[b]	2.006	1	.157		
Likelihood Ratio	4.162	1	.041		
Fisher's Exact Test				.103	.078
N of Valid Cases	13				

Table 14. Chi-Square Tests

Inferences:

- Since the Pearson Chi-Square value for Test 1 (=0.048) is less than 0.05, hence null hypothesis is rejected and the alternate hypothesis is accepted as given in TABLE 4.28
- Therefore, it is proved that their a relation between rain water harvesting facility available or installed and time required for the water logging to subside.

5 CONCLUSIONS

The monsoon in India is the reason for India to be such a prosperous country. It is the source of huge amount of fresh water which enables farming in the country and supplies the population with fresh water to cater to all its needs. However, if this is not properly managed or the natural process is disturbed due to human activities, it can disturb the environmental balance and cause lots of problem to the human population.

Pune is a rapidly developing city with the population growing exponentially in the last decade. Owing to its location and weather, it has great potential to be a one of the major cities of the country. This has led to rapid urbanization and immense construction activities. This has led to increase in concrete cover and reduction in percolation area. As a result, less water is able to penetrate the ground and reach the underground aquifers. This has also increased the amount of runoff generated. With runoff increasing but the storm water handling capacity being limited and constant, it has led to increased water logging in the city. Due to lower water percolation, the city is now facing water crisis every summer.

A solution to these problems is installing Rain Water Harvesting (RWH) systems. This enables rain water to be stored and either used directly or used to recharge the underground water tables. This ensures better utilization of the rain water. This also generates lesser runoffs and thus reduces the pressure on the drainage systems.

From the results obtained from tests conducted based on data collected from the people living across various locations in Pune using survey, we have been able to establish the fact that availability of rain water harvesting systems has direct impact on water logging problems and also ensures quick removal of accumulated water.

We have also observed that most of the housing societies do not have rain water harvesting systems installed. Even though the local laws mention such facilities to be installed, no proper monitoring system is in place. This has resulted in many societies to have been developed without any RWH facility installed. Even if installed, they are not properly maintained which causes their efficiency to reduce and eventually become non-functional. Strict laws with proper monitoring system and periodic audit is necessary for proper implementation of this strategy. Fines or penalties should be levied for non-compliance with such laws. Stringent measures will ensure efficient functioning of the system and the entire population can reap the benefits.

With the rapidly growing population and increasing pressure on natural resources, rain water harvesting is an efficient solution to all the water problems. Proper maintenance and implementation will ensure reaping of maximum benefits from this facility.

5.1 Remedial Measures

Authorities should make some new strategies and follow it such as:

- Clearing and protecting water Bodies from encroachment.
- Levying variable tax based on the usage.
- Cutting of subsidy for misusing and excess utilization of water.
- Strict enforcement of Rain water harvesting.
- Enacting strict law and enforcement for protecting water Bodies.
- Establishing more ETP and STP for treating wastewater and reuse.
- Creating water protection for awareness and acting against water theft.
- Providing subsidy or prize those who are implementing strategies for achieving sustainable water management.

IOP Conf. Series: Materials Science and Engineering 1026 (2021) 012001 doi:10.1088/1757-899X/1026/1/012001

REFERENCES

1. Magnus U. Igboekwe, Adindu Ruth, (2011), 'Groundwater Recharge through Infiltration Process: A Case Study of Umudike, Southeastern Nigeria', Journal of Water Resource and Protection, pp. 295-299.
2. S. Beecham & R. Fallahzadeh, (2011), 'Innovative approaches to urban water management in developing countries', WIT Transactions on Ecology and the Environment, Volume 150, pp. 239-247.
3. P Sundara Kumar, T Santhi, P Manoj Srivatsav, S V Sreekanth Reddy, M Anjaneya Prasad, and T V Praveen (2015), 'Storm Water Drainage Design (Case Study Vijayawada)' , International Journal of Earth Sciences and Engineering, Volume 08, No. 02, pp. 507-511.
4. Singhai, A., Das, S., Kadam, A.K. (2019),'GIS-based multi-criteria approach for identification of rainwater harvesting zones in upper Betwa sub-basin of Madhya Pradesh, India'. Environment Development and Sustainability, pp. 777–797.
5. Satheeshkumar, S., Venkateswaran, S. & Kannan, R. (2017) 'Rainfall–runoff estimation using SCS–CN and GIS approach in the Pappiredipatti watershed of the Vaniyar sub basin, South India', Modeing Earth System and Environment 3, pp. 24.
6. Etishree Agrawal, Rajat Agrawal, R.D. Garg and P.K. Garg, (2013), 'Delineation of groundwater potential zone: An AHP/ANP approach', Journal of Earth System Science 122, pp. 887–898.
7. Viavattene, Christophe & Scholes, L & Revitt, Michael & Ellis, Bryan. (2008), 'A GIS based decision support system for the implementation of Stormwater Best Management Practices'.
8. Rejani, R., Rao, K.V., Osman, (2015), 'Spatial and temporal estimation of runoff in a semi-arid microwatershed of Southern India' in Environmental Monitoring and Assessment187, pp. 54

Water Audit Study for East Tambaram

[1,] Shaik Abdul Hameed, [2]Muadh M F, [3,] Nisha Khanam

[1,2,] B.Tech Students, Civil Engg Dept, B.S.Abdur Rahman Crescent Institute of Science & Technology, Chennai, TamilNadu, India

[3,] Assistant.Prof. Civil Engg Dept, B.S.Abdur Rahman Crescent Institute of Science & Technology, Chennai TamilNadu, India

Email - shaik.hameed40@gmail.com, muadhmf98@gmail.com, nishakhanam84@gmail.com

Abstract

Water auditing isasystemto overwhelm the drought related issues, water losses, leakage etc. It plays the significant role in conservation measures to minimize the water loss which increase the efficiency of Urban Local Bodies and reduce the need to search for additional water sources. The main aim of the water audit study is to compute the amount of water that is being supplied by a water system, that is not being distributed or billed to customers. However, for the effective use of this system is to identify losses in the sums, the ULBs must first employ water auditing as a routine business practice. The various associations, such as IWA and AWWA which initiated a large-scale effort to assess the reduction of water related problems with the help of audit. Hence, a detail study on water audit is explored on the areas were water is being used in a system, identifying the sources of lost water, and an action plan is also developed to control or reduce water losses.

Keywords: ULB, IWA, AWWA, Water Audit.

Introduction

Water is virtually everything we do or use each day. India will soon be a water-stressed country and we all need to work towards our water security as the year passes. A simple action can be adopted to reduce the wastage of water and use it wisely. The study related to water shortage problem which is been existing for a long time. So, to resolve the problems a Water Auditing was done at one part of Tambaram Zone (East Tambaram). Therefore, the term water auditing is a qualitative and quantitative analysis of water consumption to identify the means of Reducing, Reusing, and Recycling the water. Some of the benefits of water auditing are to efficiently reduce water losses in the system which results in delay or avoid capital investments such as a new well, more treatment technology or additional water rights. Thus, the Urban Local Bodies takes an initiative in creating awareness among water users to understand the importance of water and take up the necessary steps to manage wastewater and save for the future. Therefore, the water audit is an effective educational and public relations tool for the water system.

Objectives of water audit

- To study and analyze the losses of water as per the IWA (International Water Association) concept by means of the population and water supply data.
- To suggest the measures to reduce the water losses for the East Tambaram.

Literatures

Amol.A.Kulkarni.et.al (2000) has determined the water balance chart as a tool for calculating the water audit, which serves as a framework for assessing a water loss situation. Water balance helps to find the recoverable losses, value the losses and to prepare the cost benefit analysis. Ganorkar (2002), has discussed about the study on water audit which covers the emphasis the importance of total water resources and their distribution. Jeff Sturman (2004) adopted the water auditing method for quantifying the water flows in easy or complicated systems, with a view to reduce water usage, as in many places people have difficulty to access the drinking water. Kedar.C & Astashil Bhabhulkar (2012) described that water auditing is a tool to overcome physical losses due to pipe leakage and overflow, which helps to identify the efficient use to reduce the capital and operating cost. Mansi (2017) focused on analysis of total water usage for the study which is found more than the capital limit. Hence, it is necessary to optimize the excess water usage. Some of the suggestions has been discussed to minimize the excess usage and water losses, and inferred that, the water meters is an essential step for optimizing the water losses and therefore it is installed to decrease the water usage limit up to 8%.

Methodology of water audit adopted

The methodology for water audit is framed in such way to classify and measure all the water uses in the system, which is called as a balance. Therefore, all uses of water in the system are equal such that the amount of water input by the sources are accounted in the standard water balance by eliminating the required , and not required quantity for the calculation of the term unaccounted for water. The most common way to view the standard water balance which is represented in terms of an separate equations or in the form of worksheet , which was developed by American Water Works Association (AWWA) and International Water Association (IWA) in 2000, were the study is carried out by the graphical standard water balance starting with the System Input category. Therefore, it is important to understand the height of the System Input category represents all the data of water pumped by the system in each time. The amount of water can be divided into two additional categories (i.e) Authorized Use and Water Losses. Therefore, Authorized Use + Water Losses = System Input. As per the concept adopted by IWA, water balance table which holds the true water measurements, shown in the Figure.1.

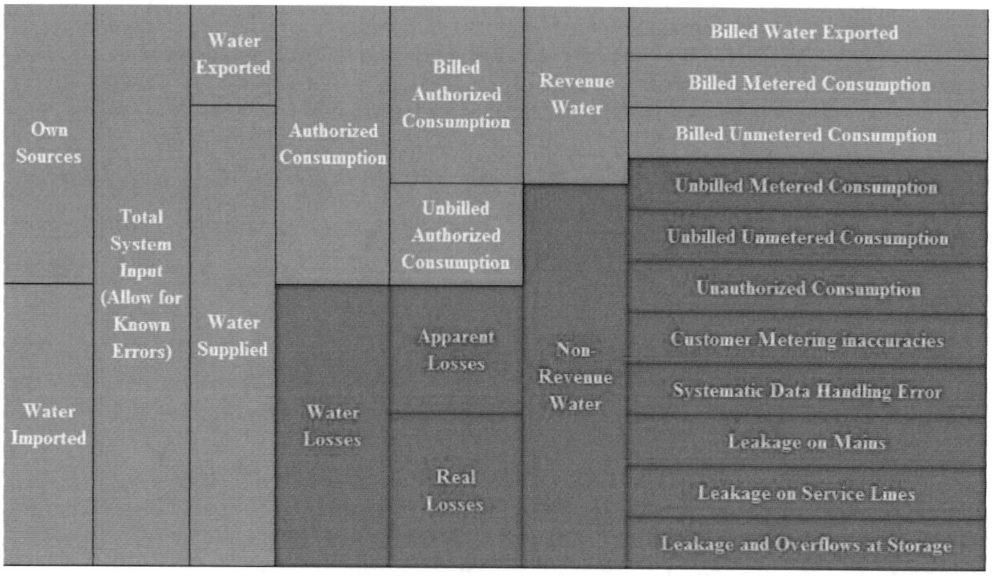

Figure 1. Water Balance Table (Source - IWA)

It is noticed from the Water Balance Table, shown in Figure.1 that some volumes are simply sums of other volumes. For instance, Authorized Consumption = Billed Authorized Consumption + Unbilled Authorized Consumption. However, Water Losses and Real Losses are deduced, or derived, volumes—in other words, volumes that are calculated through a process of elimination:

- Water Losses = Authorized Consumption - Water Supplied.
- Real Losses = Apparent Losses - Water Losses.

Since Water Losses and Real Losses are derived volumes, they are not entered by the auditor. Water balance is calculated accurately by determining the volume of Water Losses and Real Losses.

Salient details of study area

Tambaram municipality meets the water supply-demand through surface and subsurface sources. The local body-initiated water supply to the town from Palar River which is situated about 40 km which is augmented to the study area through TamilNadu water supply and drainage board. Water collected at Villiyambakkam head works are diverted to the newly constructed headworks at Melachery. The scheme was completed in 2010 and since then the entire quantity of water to Tambaram is supplied from the newly constructed Melachery Headworks. During the time of water scarcity, the water is supplied through lorries to meet the demands. The groundwater is used only during emergency as the groundwater recharge takes more times. Hence, The Figure.2 represents the distribution of water for the residents through various modes.

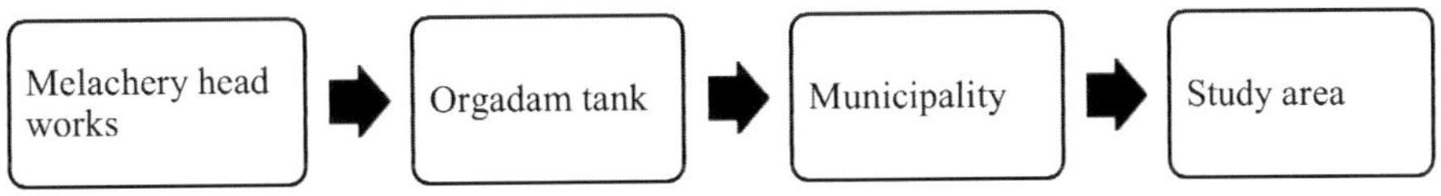

Figure 2. Distribution network of Water Supply to study area

Table 1. Details of study area

S.No	Name/ Location of Service Reservoir	Storage Capacity (LL)	Ward Covered
1	Muthurangam Park	3.64	5,6,7(P),8(P),9(P),10, 29,30(P),33(P),34,35, 36,37(P),38(P)
2	Devanesan Park (GLSR)	3.00	33(P),37(P)
3	Konekrishna Nagar	3.00	30(P),31,33(P),38(P)
4	Thiruneermalai road	3.00	1,8(P),9(P),39
5	CTO colony	4.00	32(p)
6	Krishna Nagar	3.00	32(p)
7	Thiru.Vi.Ka Nagar	3.00	2,3,4,7(P)
8	Tamil park	6.00	28,29(P),34(P)
9	Gandhi park	3.60	11,12,14,15,16(P),17(P), 23(P),24(P),25,26(P)
10	Subbarayan park	12.00	13,16(P),17(P),18(P), 19(P),22(P),23(P)
11	Srinivasa Nagar	5.00	19,20(P),21
12	Barathi Nagar	3.00	18(P),19(P),20(P),22(P)
13	Ganesh Nagar	3.00	18(P),19(P)
14	Irumbuliyur	2.00	24(P),26(P),27

Table.1 represents the areas covered for the study are been highlighted.

Field Test Analysis

Tambaram east consists of 6 water supply zones in that we randomly selected one water supply zone for conducting the field test, where few houses were selected. The time taken for the pipeline to fill the 5 litre water can at the time of water supply in the selected zone was noted. We selected 1 water supply zone at Irumbuliyur OHT for the field test, which has a tank capacity 0.2MLD, Water supplied on the day of d-system 0.2 MLD , and the house service connection at that particular zone are 188, for which the total consumption is about 1.55 LL. Samewise, the time taken to fill the 5 litres of can, during the water supply of 1 hour 30 min, is nearly about 18.48 sec. From, the above process we can also calculate the consumption of water, losses, and supply. Finally, the outcome of water audit in the water balance table were calculated.

Suggestive measures for the existing water supply system

- Overflows in Sumps should be avoided by proper scheduling of pump set operations.
- Visible leakages in Water Treatment Plant should be arrested immediately.
- Visible leakages in Transmission mains, Feeder mains, Distribution system and Valves should be arrested immediately.
- Repair in ELSRs should be done immediately for arresting minor leaks.
- Leaks in service line must be attended immediately.
- Public awareness should be created to close the tap in HSC as soon as the required water is collected to avoid wastage.
- Maintenance personnel may be educated to write the logbook of pump sets in the standard format to monitor the quantity of water pumped at each location, water received and delivered in each ELSR.
- Proper reporting of pumping return should be established to monitor the water supply at least on daily basis.

Finally, the bulk flow meters, and pressure gauges should be installed at all pump Stations, Sumps, ELSRs and critical points in the distribution network to measure flow and pressure in the lines

Conclusion

From the water audit study, it is concluded that the utilization of water should be used based on the requirement and the need without any losses and wastage. During the earlier period, the consumers has dealt with user-oriented water conservation practices and devices. The devices used was quite effective in reducing consumption and also often results in various additional benefits such as water utility which includes increase in revenue to the utility by introducing lumps within uses of various sectors, giving reduction rate on reduced consumption & giving grants for reuse of waste water use and rainwater harvesting.

Notations

ULBs – Urban Local Bodies
IWA - International Water Association
AWWA – American Water Work Association
LL – Lakh Litres
sec – Seconds
OHT – Overhead Tank
MLD - Million Litre per Day
HSC – House Service Connection
ELSR – Elevated Service Reservoir

References

1. Amol Kulkarni.A. 2000, –Water Audit: A case study of water supply scheme of Shrivardhan‖, Journal of Computing Technologies Vol. 3 Issue 6, pp. 2278 – 3814.
2. Ganorkar.R.A (2002),‖Water Audit- A Tool for Assessment Of Water Losses‖, published by International Journal Of Computational Engineering Research (ijceronline.com) Vol. 3 Issue. 3.
3. Jeff Sturman (2004), –Water auditing: A case for statutory requirements‖, Water Intelligence Online, Vol 8, Issue 6.
4. Kedar Chimote (2012), –Water Audit‖, National Conference on Innovative Paradigms in Engineering & Technology (NCIPET-2012) Proceedings published by International Journal of Computer Applications (IJC), pp.17-20.
5. Mansi (2017) –Water Audit and inevitability of Water Meter‖, International Research Journal of Engineering and Technology (IRJET), vol 4, Issue 4.

IOP Conf. Series: Materials Science and Engineering 1026 (2021) 012003 doi:10.1088/1757-899X/1026/1/012003

Investigational study on Influence of Fiber Reinforced Polymer Wrapping on Concentrically Loaded Concrete Column

NPannirselvam[1], BSudalaiGunaseelan[2], JRajprasad[3]
[1]Associate Professor, Department of Civil Engineering,
[2]Assistant Professor, Department of Civil Engineering,
[3]Assistant Professor, Department of Civil Engineering,
SRM Institute of Science and Technology, Potheri, Kattankulathur - 603203,
Chengalpattu District, Tamilnadu, India.
Corresponding author Email: [1]pannirsn@srmist.edu.in,

Abstract. The aim of research work to study the column with varying dimension and parameter. Along with wrapping of the column. The study has been made on axial load, axial deflection and lateral deflection. There 18 specimens have been casted and testing is done. Such as 300mm height and 150mm diameter column of six numbers, 450mm height and 150mm diameter columns of six numbers, 600mm height and 150mm diameter columns of six numbers. The 300mm height of the columns has been fully wrapped with 0-degree orientation of one-layer thickness, fully wrapped with 45-degree orientation of one-layer thickness, partially wrapped with 0-degree orientation of double-layers thickness, partially wrapped with 45-degree orientation of double layers thickness, fully wrapped with 45-degree orientation of double layers thickness and one number of unwrapped concrete columns. The 450mm height of the columns has been fully wrapped with 0-degree orientation of one-layer thickness, fully wrapped with 45-degree orientation of one-layer thickness, partially wrapped with 0-degree orientation of double-layers thickness, partially wrapped with 45-degree orientation of double layers thickness, fully wrapped with 45-degree orientation of double layers thickness and one number of unwrapped concrete column. The 600mm height of the columns has been fully wrapped with 0-degree orientation of one-layer thickness, fully wrapped with 45-degree orientation of one-layer thickness, partially wrapped with 0-degree orientation of double-layers thickness, partially wrapped with 45-degree orientation of double layers thickness, fully wrapped with 45-degree orientation of double layers thickness and one number of unwrapped concrete column. The specimens have been tested and corresponding results are obtained.

Key words: Column, Compressive strength, Glass fiber reinforced polymer, Reinforced concrete, Wrapping.

1. Introduction

Concrete is the most versatile material used in the construction activities. Repair and rehabilitation of structure evolved to retain the monument and olden structures. In certain cases the newly constructed structures also fails its ultimate load and this emerged the way to Fiber Reinforced Polymer (FRP). FRP used not only to strengthen the existing structure and in order to diminish the size of structural components and it favours. FRP material possess high strength to weight ratio and does not corrode.

During earthquake the confinement of the FRP materials provides allowable time to escape from its natural calamities. In later stage after earth quake occur and FRP used to reclaim the strength of affected concrete structures. The types of FRP available in the market are carbon, glass and aramid fibers whereas reduction of cost only in the glassfiber. Various research investigators choose the fiber based on their applications and not by the cost of fiber. Recently basalt fibers from melted natural basalt rock and performs tensile strength, corrosion resistance and fire.

2. Objectives

In this research article the performance of reinforced concrete (RC) columns wrapped with glass fiber reinforced polymer (GFRP) sheets was investigated. Fifteen RC columns were fabricated. Two types of orientation and different types of dimensions of concrete columns. The compressive behaviour of the GFRP-confined reinforced concrete columns was studied in details, including the axial load, axial deflection and lateral deflection.

3. Experimental Program

Circular RC short columns of grade M40 of 300mm height 150mm diameter of 6 specimens, 450mm height 150mm diameter of 6 specimens and 600mm height 150mm diameter of 6 specimens designed as per IS 456:2000. Initially each one specimen is tested without wrapping and different orientations due to wrapped specimens are tested.

4. Materials

4.1 Cement

The properties of cement shown in Table 1.

Table 1. Portland pozzolana cement

S.No.	Properties	Test results
1	Specific gravity	3.10
2	Fineness test	3%
3	Consistency test	31%
4	Initial setting time	40 minutes
5	Final setting time	310 minutes

4.2 Fine aggregate

Fine aggregate confirming to Zone II as per IS: 383-1970 with specific gravity and fineness modulus were 2.51 and 2.64 respectively.

4.3 Coarse aggregate

Gravel of size 10-12mm size 40% and 20mm size 60% for M40 mix was used as coarse aggregate. Its specific gravity and fineness modulus were 2.78and 6.87 respectively.

4.4 Admixture

Conplast SP430 admixture based on sulphonated naphthalene polymers.

4.5 Steel

Reinforcing steel of 8mm diameter was used as stirrups and 12mm diameter was used as main reinforcement and 6 numbers of reinforced are used. Table 2 shows the mechanical properties of steel.

Table 2. Mechanical properties of reinforcing steel

S.No.	Diameter, mm	Yield stress, f_y (MPa)	Yield strain (ε_y)	Elastic modulus, E (MPa)
1	8	455	0.0024	210060
2	12	447	0.0024	210060

5. Mix Proportions

The mix proportioned for M40 presented in Table 3.

Table 3. Mixture proportion ofM40

S.No.	Mixture ID	M40(CC)
1	Cement (kg/m^3)	430
2	Fine Aggregate (kg/m^3)	542
3	Coarse Aggregates (kg/m^3)	1023
4	Water (kg/m^3)	159
5	W/C ratio	0.37
6	Admixture (liter/m^3)	3.44

6. Results and Discussion

6.1 Compressive test for 1:3 cement mortar

Casting of cement mortar is 1:3 ratio and size of the cubes is 70.6mm compressive test are carried out cube's specimens in compression testing machine. Cubes were tested 3 days and 7 days of casting. Figure.1 testing of cement mortar respectively in compression testing machine.

Figure 1. Testing of cement mortar

Table 4. Mortar test

S.No.	Duration Days	Compressive Strength N/mm^2

1	7	27.66
2	14	38.68
3	28	59.68

6.2 Compressive strength results for concrete

The test for compressive strength was done on standard cubic specimens (150X150X150) mm of M40 mix, after curing for 3, 7 and 28 days according to IS 516. 1000kN capacity compression testing machine tested for the cubes which placed precisely in the center of the spherically seated upper block of the loading head and the compressive load was given at a rate of 0.2kN/s. The values were documented and shown in table 5.

Figure 2. Compressive test

Table 5. Compressive strength results

S.No.	Tested specimens Days	Compressive Strength N/mm^2
1	7	28.46
2	14	38.02
3	28	45.26

6.3 Split tensile strength tests

The split tensile test is performed for conventional concrete specimens and presented in figure.3.

Figure 3. Split tensile test

Table 6. Split tensile strength

S.No	Split tensile strength N/mm^2
1	6.07
2	5.73
3	5.49

6.4 Axial load test for GFRP wrapped and unwrapped RC columns

Wrapping of reinforced concrete column is used different types of orientations like 0-gedree orientation and 45-degree orientation. GFRP due to wrapped the specimens are partially wrapped and fully wrapped. Partially wrapping on specimens is 0 and 45 degree orientation was wrapped 2-layers thickness. Fully wrapping on specimens is 0 and 45 degree orientation was wrapped 1-layer thickness and another one specimen are wrapped 45-degree orientation used 2-layers thickness. GFRP wrapping on specimens is used different types of specimens 300mm, 450mm and 600mm height.

Figure 4. Partially and fully wrapped reinforced concrete columns with different orientations

6.5 Axial deflection of GFRP wrapped and unwrapped 300mm height RC column specimens

Reinforced concrete column with fully and partially wrapping on 0 and 45 degree orientation are wrapped1-layer thickness and 2-layers thickness of 300mm height. Figure 5 shows the load Vs axial deflection with fully wrapped, partially wrapped and unwrapped reinforced concrete columns.

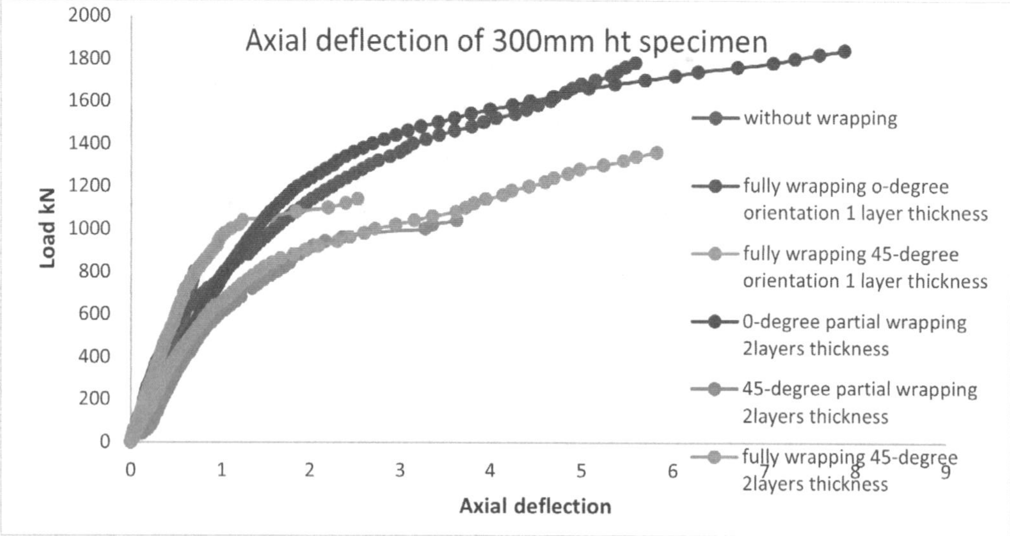

Figure 5. Axial deflection of GFRP wrapped and unwrapped RC columns

6.6 Lateral deflection of GFRP wrapped and unwrapped 300mm height RC column specimens

Reinforced concrete column with fully and partially wrapping on 0 and 45 degree orientation are wrapped 1-layer and 2-layers thickness of 300mm height. Figure 6 shows the load Vs lateral deflection with fully and partially wrapped reinforced concrete column.

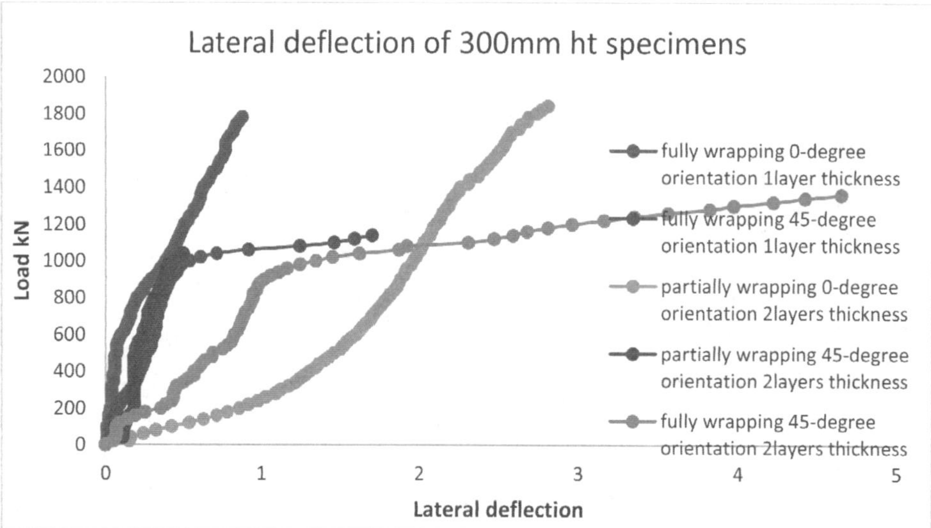

Figure 6. Lateral deflection of GFRP wrapped RC columns

6.7 Axial deflection of GFRP wrapped and unwrapped 450mm height RC column specimens

Reinforced concrete column with fully and partially wrapping on 0 and 45 degree orientation are wrapped 1-layer thickness and 2-layers thickness of450mm height. Figure 7 shows the load Vs axial deflection with fully wrapped, partially wrapped and unwrapped reinforced concrete columns.

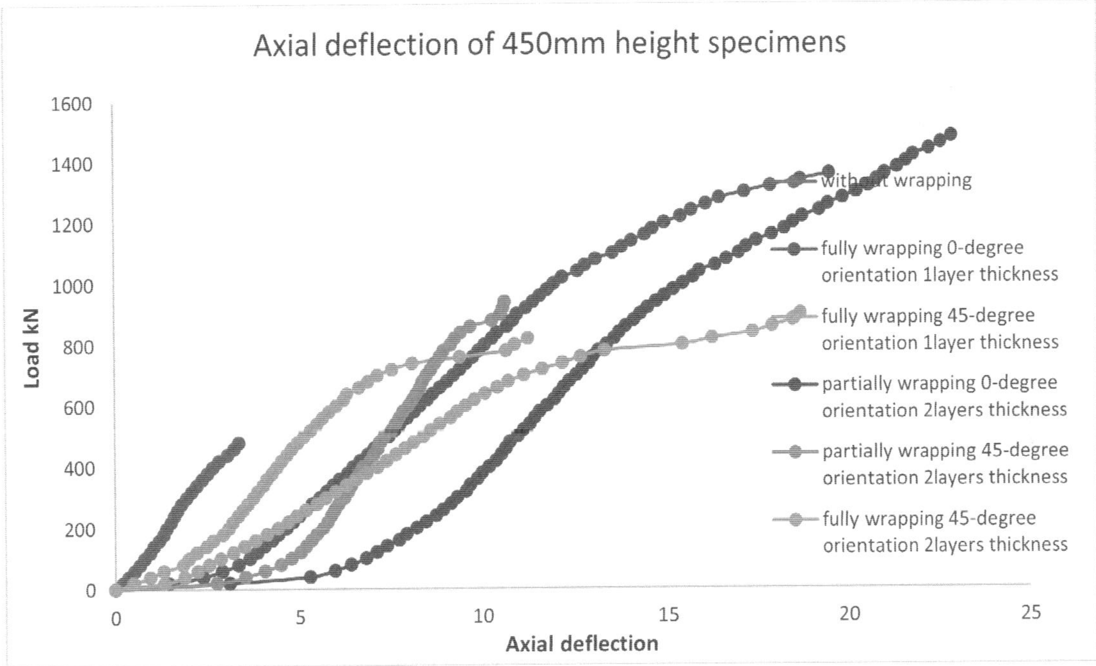

Figure 7. Axial deflection of GFRP wrapped and unwrapped RC columns.

6.8 Lateral deflection of GFRP wrapped and unwrapped 450mm height RC column specimens

Reinforced concrete column with fully and partially wrapping on 0 and 45 degree orientation are wrapped 1-layer thickness and 2-layers thickness of450mm height. Figure 8 shows the load Vs Lateral deflection with fully and partially wrapped reinforced concrete column.

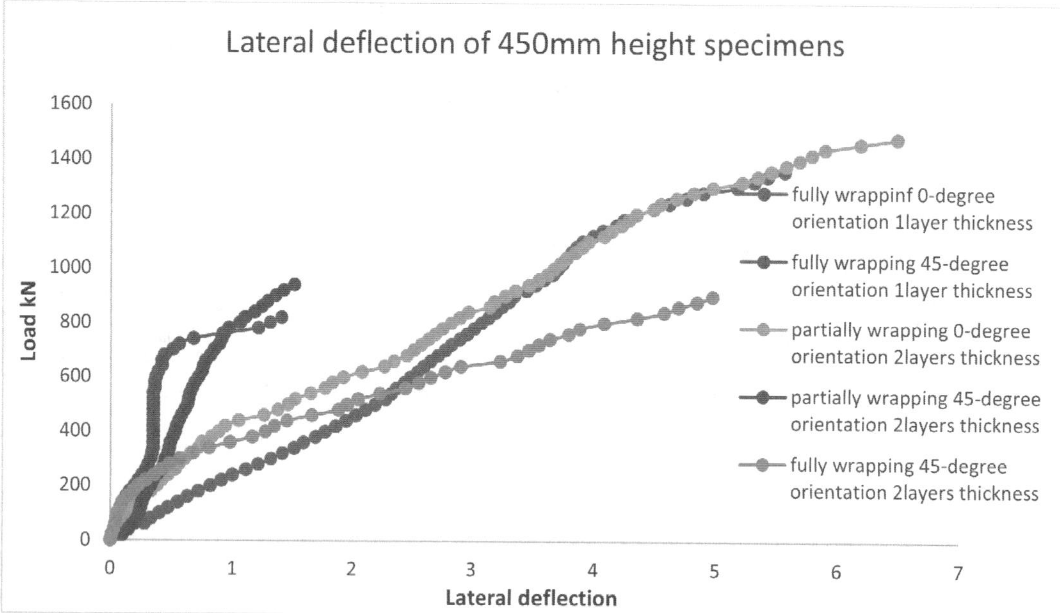

Figure 8. Lateral deflection of GFRP wrapped RC columns.

6.9 Axial deflection of GFRP wrapped and unwrapped 600mm height RC column specimens

Reinforced concrete column with fully and partially wrapping on 0 and 45 degree orientation are wrapped 1-layer thickness and 2-layers thickness of600mm height. Figure 9 shows the load Vs axial deflection with fully wrapped, partially wrapped and unwrapped reinforced concrete columns.

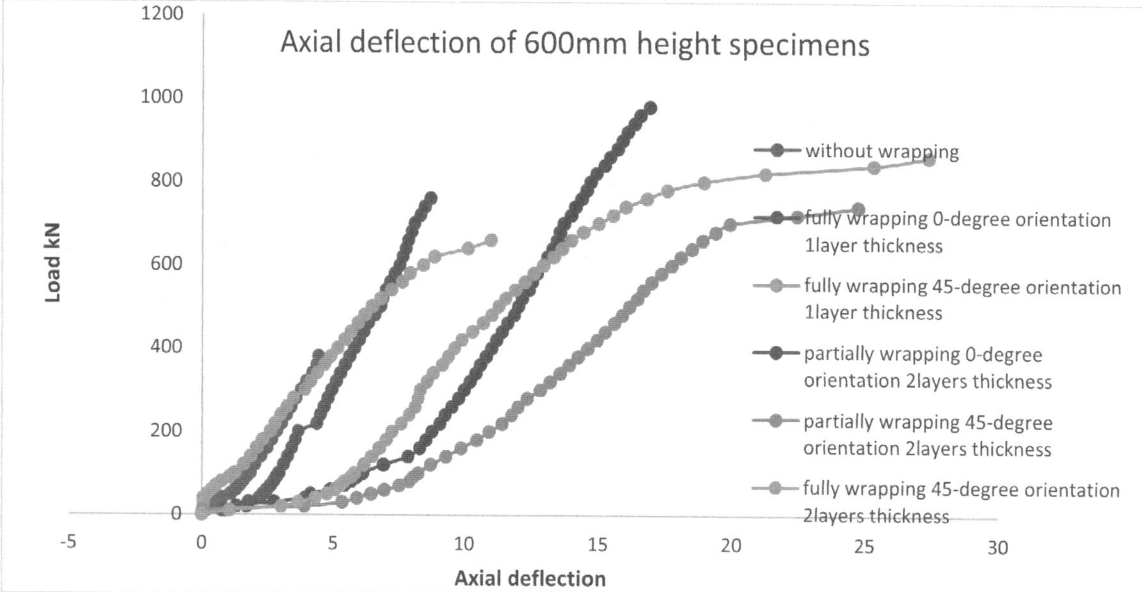

Figure 9. Axial deflection of GFRP wrapped and unwrapped RC columns.

6.10 Lateral deflection of GFRP wrapped and unwrapped 600mm height RC column specimens

Reinforced concrete column with fully and partially wrapping on 0 and 45 degree orientation are wrapped 1-layer thickness and 2-layers thickness of600mm height. Figure 10 shows the load Vs Lateral deflection with fully wrapped and partially wrapped reinforced concrete column.

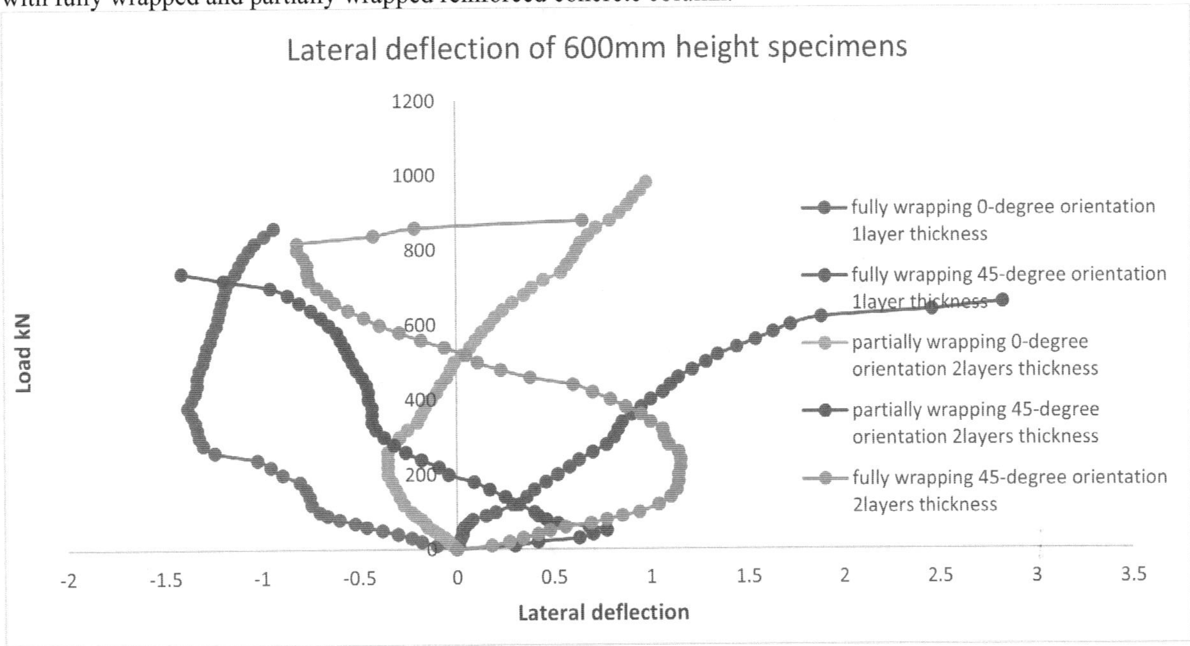

Figure 10. Lateral deflection of GFRP wrapped RC columns.

6.11 Axial load strength of various types of reinforced concrete columns

Figure 11. Unwrapped, fully and partially GFRP wrapped RC columns.

Table 7. Axial load strength of various types of RC columns.

S NO.	Different types of GFRP Wrapping	300 mm height 150 mm diameter	450 mm height 150 mm diameter	600 mm height 150 mm diameter
1	Without wrapping of reinforced concrete columns	787kN	447kN	343kN
2	Fully wrapped 0-degree orientation 1-layer thickness	1728kN	1298kN	817kN
3	Fully wrapped 45-degree orientation 1-layer thickness	1074kN	757kN	629kN
4	Partially wrapped 0-degree orientation 2-layer thickness	1800kN	1410kN	963kN
5	Partially wrapped 45-degree orientation 2-layer thickness	993kN	870kN	711kN

6	Fully wrapped 45-degree orientation 2-layers thickness	1308kN	819kN	829kN

6.12Axial load strength of wrapped and unwrapped concrete columns 300mm height and 150mm diameter

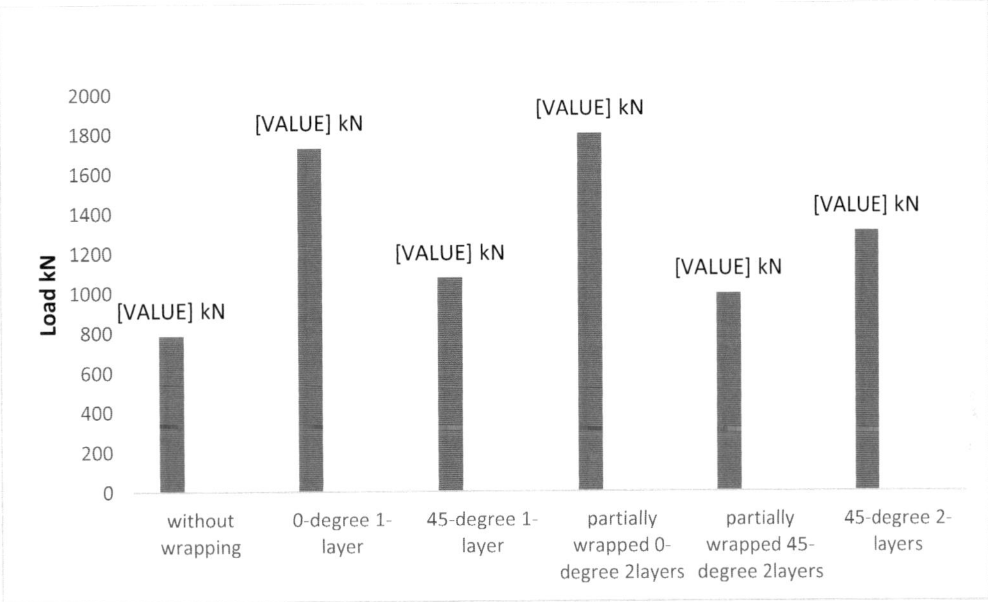

Figure 12. Axial load strength for 300mm height RC columns.

Axial load strength of wrapped and unwrapped concrete columns 450mm height and 150mm diameter

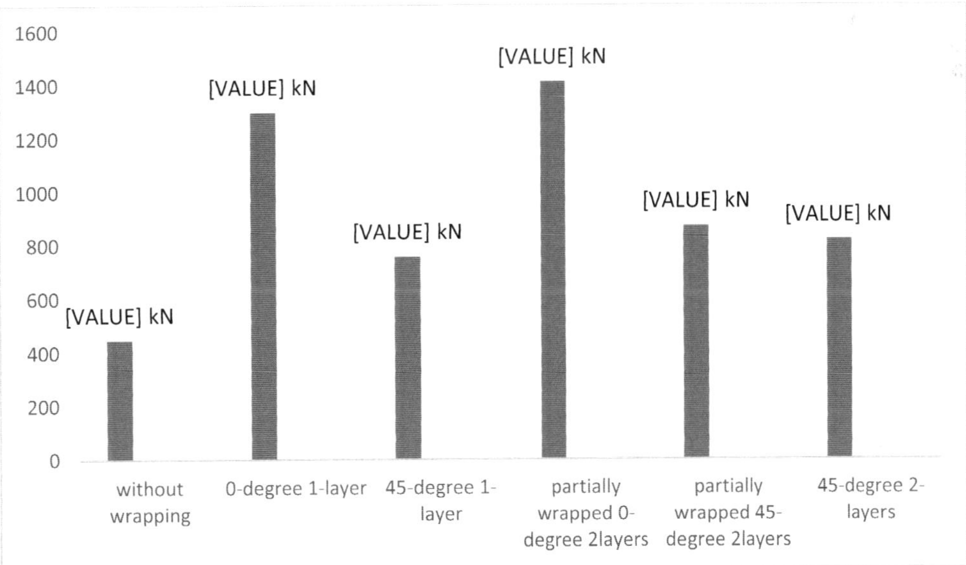

Figure 13. Axial load strength for 450mm height RC columns.

Axial load strength of wrapped and unwrapped concrete columns 600mm height and 150mm diameter

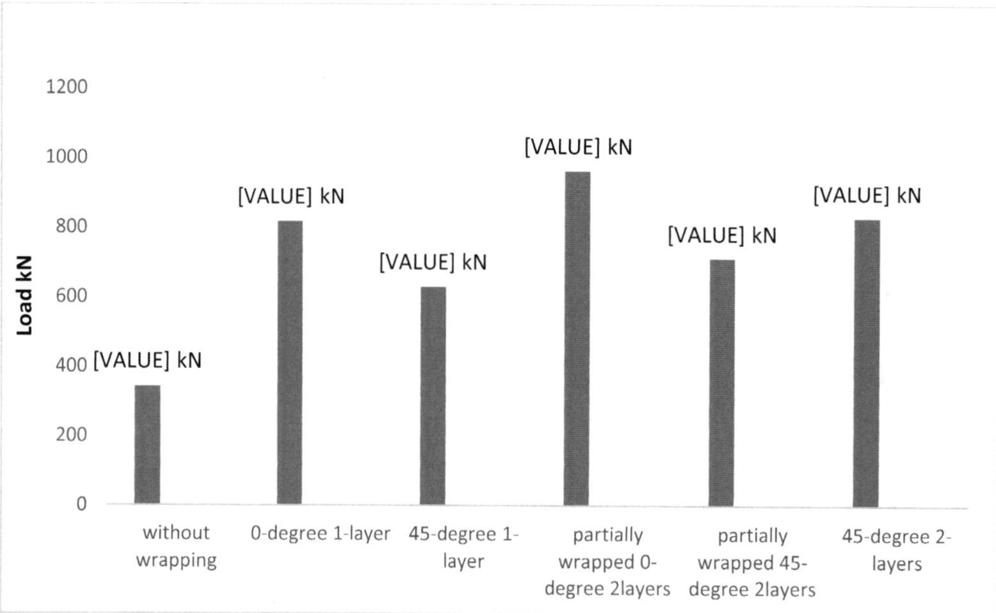

Figure 14. Axial load strength for 600mm height RC columns.

7. Conclusion

1. Reinforced concrete columns without wrapping was compared to fully wrapped 0 and 45-degree orientation of 1-layer thickness of wrapped concrete column strength achieved 220 and 135%.
2. Partially wrapped 0-degree orientation of double layer thickness of wrapping column reached high strength compared to fully wrap single layer of 0-degree orientation of reinforced concrete columns.
3. Without wrapping of reinforced concrete column strength is compared to fully wrapped 45-degree orientation of double layer thickness of wrapped concrete column strength is above 165%.
4. Partially wrapping of 0 and 45-degree orientation of double layers thickness of concrete column attained 230 and 125% when compare to without wrapping of reinforced concrete columns.

8. References

[1] Kunal R BhoiDr. Kishore Ravande and Abhijeet Galatage 2019 Strengthening of concrete by GFRP and CFRP wrapping International advanced research journal in science, **6**69-73.

[2] HananSuliman Al Nimry and Rabi Amer Al Rabadi 2019 Axial flexural interaction in FRP wrapped RC columns International journal of concrete structures and materials1-19.

[3] Ruqayyah Ismail, Raizal S.M Rashid, Wong chienchan, MohdsalehJaafer and FarzadHejazi 2018Compressive behaviour of concrete cylinder fully and partially confined by carbon fiber reinforced polymer *Constr. Build. Mater.* **201**196-206.

[4] Jun Jie Zeng, Yong Chang Guo, Wan Yang Gao and Wei Peng Chen, Li Juan Li 2018 Stress strain behaviour of concrete in circular concrete columns partially wrapped with FRP stripsComposite structures**200** 810-28.

[5] Sahar Y Ghanem and Issam E Harik2018Concentrically loaded circular RC columns partially confined with FRP International journal of concrete structures and materials1-17.

[6] Thong M. Pham, Jim Youssed, Muhammad NS Hadi and Tung M Tran 2016Effect of different FRP wrapping arrangement on the confinement mechanism*Procedia Eng.* **142** 307-13.

[7] Thong M. Pham, Muhammad N. S. Hadi and Jim Youssef 2015Optimized FRP schemes for circular concrete columns under axial compression *J. Compos. Constr.* **19**1-18.

[8] Azadeh Parvin and David Brighton 2014Review on FRP composites strengthening of concrete columns under various loading conditions Polymers**6**1040-56.

[9] Faizan Tahir M, Yaqub M, Imran Bukhari, Usman Arshid M and Rameez Sohail M 2013Effect of CFRP confinement on plain and reinforced concrete square columns Life science journals **10**770 -7.

[10] Kinjal V Ranolia, B K Thakkar and J D Rathod2013Effect of different patterns and cracking in FRP wrapping on compressive strength of confined concrete*Procedia Eng.* **51** 169-75.

[11] Rahul Ravala andUrmilDaveb2013Behavior of GFRP wrapped RC columns of different shapes, International conference on engineering*Procedia Eng.* **51** 240 – 49.

[12] Muhammad N S Hadi and Ida bagulraiwidiarsa2012Axial and flexural performance of square RC columns wrapped with CFRP under eccentric loading*J. Compos. Constr.* 640-49.

[13] Hadi M N S and Yazici V 2010Behaviour of FRP wrapping circular reinforced concrete columnsChallenges opportunities and solutions in structural engineering and construction 215-20.

[14] R Eid, N Roy and P Paultre2009Normal and high strength concrete circular elements wrapped with FRP composites *J. Compos. Constr.* **13**113-124.

[15] Kumutha R, Vaidyanathan R and Palanichamy M S 2007Behaviour of reinforced concrete rectangular columns strengthened using GFRP *Cem Concr Compos* **29** 609-15.

IOP Conf. Series: Materials Science and Engineering 1026 (2021) 012004 doi:10.1088/1757-899X/1026/1/012004

Impact of the Lead Rubber Base Isolators on Reinforced Concrete Building

Ravi Kant[1,*], Wesam Al Agha[2], Mohindra SinghThakur[1], and N Umamaheswari[2]

[1] School of Mechanical and Civil Engineering, Faculty of Engineering and Technology, Shoolini University of Biotechnology and Management, Solan-Oachghat-Kumarhatti Dist., Himachal Pradesh, 173229, India

[2] Department of Civil Engineering, Faculty of Engineering and Technology, SRM Institute Of Science and Technology, Kattankulature, Kancheepuram Dist., Tamilnadu, 603203, India

* E-mail: ravikantthakur1995@gmail.com

Abstract. Civil engineering infrastructures are prone to natural disasters like earthquakes, floods, and hurricanes. Dissipation energy plays the main role in design of structures under Earthquake loads and controls the behavior of buildings when it bears the lateral loads and transfers them to the foundation. To resist the vibrations in the structure, it is important to use a high technology like Base isolation which reduces base frequency of building vibration and reaching it to a lower value. In other words, seismic isolation has the main objective to protect structure during earthquake events. Base isolation is a flexible layer between foundation and superstructure thus decoupling the structure from damaging effect of ground motion. This edits the dynamic characteristics of the structural system. A comparative study is conducted of seismic performance on three types of thickness (100mm,200mm, 300mm) in which M1T1, M2T1, M3T1, M1T2, M2T2, M3T2, M1T3, M2T3, M3T3, M1T4, M2T4, and M3T4 has been analyzed of isolators namely Lead Rubber Base Isolators (LRBI). For this evaluation, an 8-story structure has been analyzed by using (ETABS) finite element program where Models have ordinary RC moment- resisting frames with square column cross- sections. Parameters like drift, displacement, story acceleration, frequency, and story velocity have been analyzed and it has been found that due to the presence of Lead Rubber Isolators the story drift, story accelerations, base shear, lateral shear forces, and velocity are greatly reduced and story displacement is increased in both X and Y directions compared for various thickness of lead rubber base isolation systems under near and far-field earthquakes. Most importantly it has been established that the transmission of the high energy seismic waves to the superstructure is greatly reduced by introducing base isolators in between the top of the footing and the bottom of the superstructure. The base isolator with thickness 100mm for the concrete grade M20, M30, M40 performed more efficiently than other thicknesses. In conclusion, the deformation and transmissibility ratio must be minimized to introduce as a solution for absorbing the vibrations where the engineer should consider this effect for adequate design.

1. Introduction

Strong earthquakes in the past have resulted in significant property damage and loss of lives. Earthquake damage is mainly associated with damage to seismically deficient buildings, often designed prior to the enactment of modern building codes. Therefore, researchers have focused their

IOP Conf. Series: Materials Science and Engineering 1026 (2021) 012004 doi:10.1088/1757-899X/1026/1/012004

efforts on developing new and effective seismic risk mitigation strategies for such buildings. Conventional seismic protection methods, such as enlarging structural elements or adding new strengthening and stiffening elements focus on improving the capacity of buildings to meet likely demands. However, stiffening buildings results in shorter vibration periods and higher floor accelerations, resulting in higher structural damage. Therefore, a preferred approach for seismic retrofitting buildings may be to decrease demands rather than to increase capacities. This can be achieved by introducing flexible devices to buildings at their ground levels to isolate them from the transfer of seismic energy. Research aims to improve the seismic performance of strategically important buildings such as commercial buildings, industrial structures, elevated water tanks, residential houses, etc. in addition to places where sensitive equipment are intended to protect from the hazardous effects during earthquake, base isolation is recognized as a mature and efficient technology that can be adopted to reduce the momentum of the seismic waves. The lead rubber base isolator dissipates a larger quantity of seismic energy as the lead plugs experience large hysteresis deformation. In addition, the lead rubber base isolators have a much higher lateral stiffness. This means that the incorporation of lead rubber base isolators can contribute towards lateral resistance to non-seismic loads, thereby controlling base displacements during strong earthquakes. The use of the isolation system depends on the extent of control to achieve the better for seismic response and thereupon its design is done to suit the requirement for structure. To design the base-isolated structures, the superstructure is decoupled from earthquake ground motion by introducing a flexible interface between the foundation and the base of the structure. Therefore the fundamental time-period of the structure is shifted to a large value for absorbing the energy in damping, limiting the amount of force that can be transferred to the superstructure such that inter-story drift and floor accelerations are decreased drastically. It also consequently avoided the matching of fundamental frequencies of base-isolated structures and the predominant frequency contents of earthquakes, leading to a flexible structural system with ductility and more suitable from earthquake resistance viewpoint. To present the lead rubber base isolator as shown in figure 1 and figure 2.

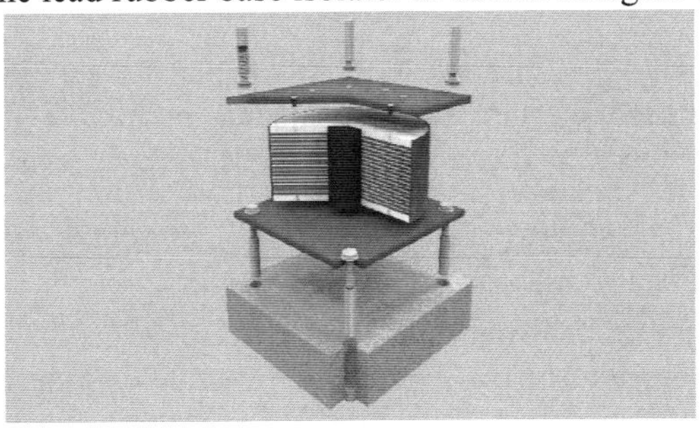

Figure 1. LRB seismic isolator

Courtesy: https://doshinrubber.com/lead-rubber-bearing/

Figure 2. LRB seismic isolator

Courtesy: http://welllink.com.tw/en/lead-rubber-bearing

2. Literature survey

Different sources of damping that are considered depending on the nature of the building. Furthermore, many types of isolators are studied, in which lead rubber base isolators were deeply enlightened with its importance.

2.1. Ras, I. Nait Zerrad (2016)

A 3D analytical study was done by taking the seismic behaviour of a 12-storey steel building retrofitted with a FPS. Fast nonlinear time history analysis (FNA) of Boumerdes earthquake (Algeria, May 2003) was conducted for analysis and carried out using SAP2000 finite element program. To compare the structural performance with and without this tool of dissipation energy, the results of models were discussed and concluded to show the interesting potential of the FPS isolator out using SAP2000 finite element program. To compare the structural performance with and without this tool of dissipation energy, the results of models were discussed and concluded to show the interesting potential of the FPS isolator. [3]

2.2. *Mital N. Desai (2015)*

It was studied the comparative study of performance by using three types of base isolators namely Lead Rubber Bearing (LDRB) , Low Damping Rubber Bearing (LDRB) and High Damping Rubber bearing (HDRB). Eight -storey building was analysed using the Response Spectrum Method. Dynamic analysis was done using STAAD-Pro software. Parameters like Base shear, Building displacement and frequency and spectral acceleration were compared for the building with base isolator and building with fixed base. [20]

2.3. *Minal Ashok Somwanshi (2015)*

It has developed a new base isolation system for a multi-storey RC building. This study occurred to analyze the 13-storey rigid jointed plane frame for two cases. The first case was fixed base and the second case was base isolated. Analytical study was done using ETABS software with Bhuj earthquake ground motion records. Maximum vertical reaction was included from analysis in ETABS software. Lead rubber bearings were designed manually using this vertical reaction and total mass of building. In order to evaluate floor response, Time-history analysis was conducted to have acceleration and displacement values under seismic loads. This intends to demonstrate how an isolation system can be efficient, evaluating its effectiveness for the building in terms of maximum shear force, maximum bending moment, base shear, storey drift and storey displacement reductions.[19]

2.4. *Kilaret V*

4-storey reinforced concrete building was analysed according to Eurocode 8 for seismic analysis and dynamic performance investigation. Different groups of base isolation devices were studied for evaluation. The first case was simple rubber bearing and the second one was lead rubber bearing as a base isolation system. For the evaluation of each system a soft, normal and hard rubber stiffness with different damping values were used. It performed Non-linear pushover analysis with the SAP 2000 software. It was concluded that the stiffer isolators with higher damping give smaller target base displacements as compared to softer one with lower damping. It also presents that the relative displacement of the superstructure was smaller since the softer isolators were used. It can be expected that the smallest relative displacement with the use of softer isolators. If the used isolators were too stiff it cannot protect the superstructure. [17]

3. Research Statement

In order the RCC building needs to be designed by the insufficient balance, strength, and ductility that have occurred. When the dynamic analysis under seismic load happened, it is important to introduce supplementary damping for building, the isolation systems considered include Lead rubber isolator of thickness 100mm, 200mm, 300mm.It is taken Base isolation as an opposite approach, i.e. to reduce the seismic demand instead of increasing the capacity. Controlling ground motion is impossible, but we can modify the demand on structure by preventing/reducing the motions being transferred to the structure from foundations. Moreover, the use of RCC frames is very common these days. Therefore, the problem of analyzing the RCC frames by introducing base isolators has been taken for the studies of the dynamic response of such frames. The concrete grade selected for such analysis is M20, M30, and M40. The details of material other in outs etc. are discussed in subsequent paragraphs. To know the dynamic response of the base-isolated buildings for a given soil bearing capacity, four types of building frames have been generated by using software ETABS, Version (16.2.1). The detail of such models generated is given below as shown in tables (1, 2, and 3)

Table 1. Models M20 (M1)

Type	T1	T2	T3	T4
Base	Fixed	100 mm	200 mm	300 mm

Table 2. Models M30 (M2)

Type	T1	T2	T3	T4
Base	Fixed	100 mm	200 mm	300 mm

Table 3. Models M40 (M3)

Type	T1	T2	T3	T4
Base	Fixed	100 mm	200 mm	300 mm

4. Objective of the Study

The objective of the current investigation is to assess the performance of different types of base isolation systems for seismic protection of reinforced concrete frame buildings, through dynamic analysis time history analysis and static analysis of a selected prototype building.

5. Modeling and Material Properties

5.1. Modeling

The seismic behavior of the model has been studied with ordinary RC moment- resisting frames by noting the impact of changes on seismic characteristics where lead rubber base isolators occurred. According to modeling, 8-story structure storey RC building occurred .The analysis (Standard) was done by considering the connection between columns and beams is completely fixed [8].

5.2. Design Parameters

- Regarding the study, Loads acting on the structure are shown as in table(4).

Table 4. Loads.

Dead load (DL) and Live load (LL)	1 As per IS 875(PART 1) (1987) and IS 875(PART 2)
Seismic loads (SL)	As per IS 1893(PART 1) (2000)
Earthquake Seismic zone	Zone IV
Rock/soil type	Medium
Rock and soil factor	1
Damping ratio	5%

- The Preliminary data used, as shown in table(5)

Table 5. Preliminary data.

Utility of building	Residential building
Number of stores	8
Shape of building	Rectangular with RCC Frame Structure

Grade of concrete and steel	M20, M30, M40, and Fe415
Ductility design	IS:13920-1993

- Models Characteristics the following dimensions of column and beams have been considered in all the models as shown in table (6)

Table 6. Column and beams.

Dimension of beam	380 x 420 mm
Dimension of column	500 x 500 mm
Support condition	Fixed

- Height of models is shown in table (7)

Table 7. Floor to Floor height for Models (m)

Models and Types	M1T1 M2T1 M3T1	M1T2 M2T2 M3T2	M1T3 M2T3 M3T3	M1T4 M2T4 M3T4
Base	2	0.1	0.2	0.3
1	4	3.9	3.8	3.7
2	4	4	4	4
3	4	4	4	4
4	4	4	4	4
5	4	4	4	4
6	4	4	4	4
7	4	4	4	4
8	4	4	4	4

- Characteristics of model cases are shown in table (8)

Table 8. Model cases.

Name of Models	Properties for Models
M1T2, M2T2, M3T2 M1T3, M2T3, M3T3 M1T4, M2T4, M3T4	M1T2, M2T2, M3T2 M1T3, M2T3, M3T3 M1T4, M2T4, M3T4
Dimension of beam	380*420 mm
Dimension of column	500*500mm
No. of column	240
No. of beam	392

U1 linear effective stiffness for model 2	229137.19 kN/m
U1 linear effective stiffness for model 3	114568.59 kN/m
U1 linear effective stiffness for model 4	57284.29 kN/m
U1, U2, U3 linear effective damping	0.1 kN-s/m
U2, U3 linear effective stiffness	800 kN/m
U2, U3 non-linear stiffness	2500 kN/m
U2, U3 non-linear yield strength	80 Kn
U2, U3 non-linear post-yield strength ratio	0.1

For time history, Uttarkashi earthquake is used with Record as shown in figure 3.

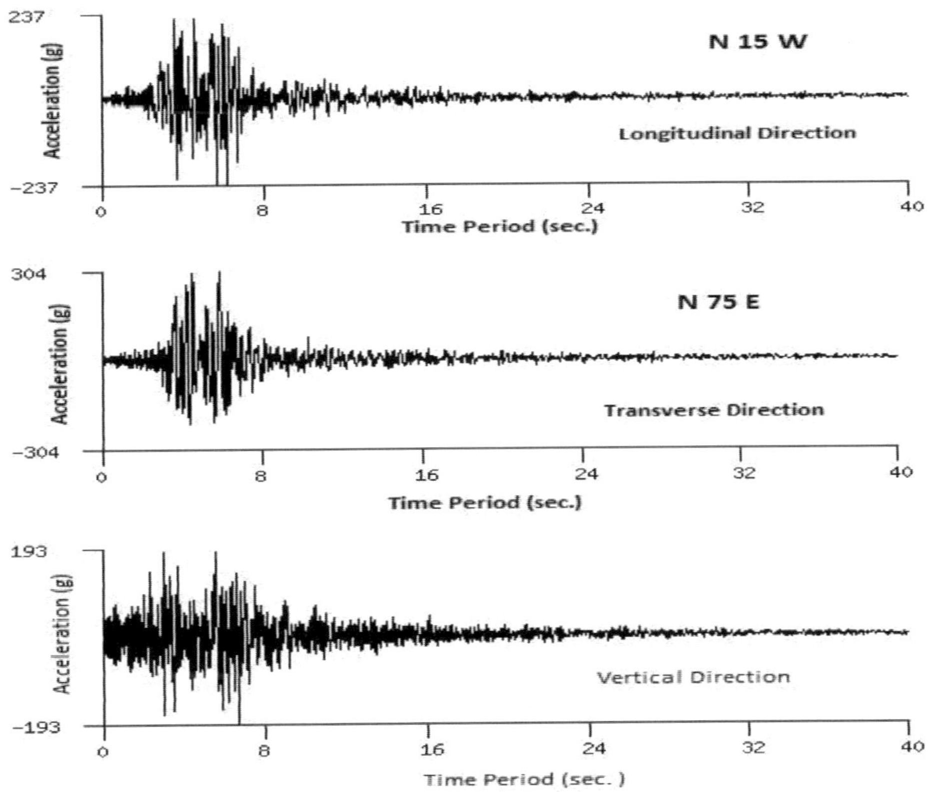

Figure 3. Uttarkashi earthquake of October 20, 1991

Direction values (Uttarkashi earthquake):

- Longitudinal (N 15 W) = 0.242 (237 cm/s/s)
- Transverse (N 75 E) = 0.310 (304 cm/s/s)
- Vertical = 0.196 (193 cm/s/s)

The procedure is generated by using ETABS with introducing Isolator to building, the procedure will not be changed for analysis methods.

6. Results and discussion

The results are carried out by ETABS, Version (16.2.1), as below to present the comparison of the dynamic performance of the structure. The response of the eight-story structure is investigated under three different thickness based on their effective damping. Parameters like drift, displacement, story acceleration, frequency, and story velocity have been analyzed and it has been found that due to the presence of Lead Rubber Isolators the story drift, story accelerations, base shear, lateral shear forces, and velocity are greatly reduced and story displacement is increased in both X and Y directions are compared for isolated building with responses of non-isolated building under near and far-field earthquake excitations.

6.1. Storey Drift

It has been observed that on the dynamic behavior of the seismic response the Story drift response decreases from ground floor to the upper floor in a building structure in both X and Y direction as shown in figure 4 and figure 5.

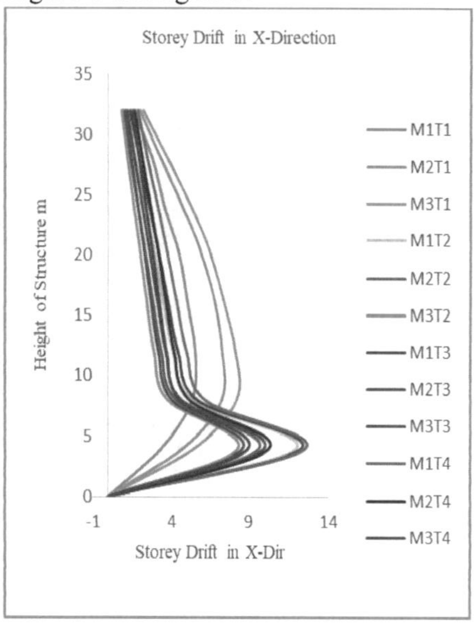

Figure 4. Comparison of story drifts in the X direction for fixed base and 100mm, 200mm, 300mm.

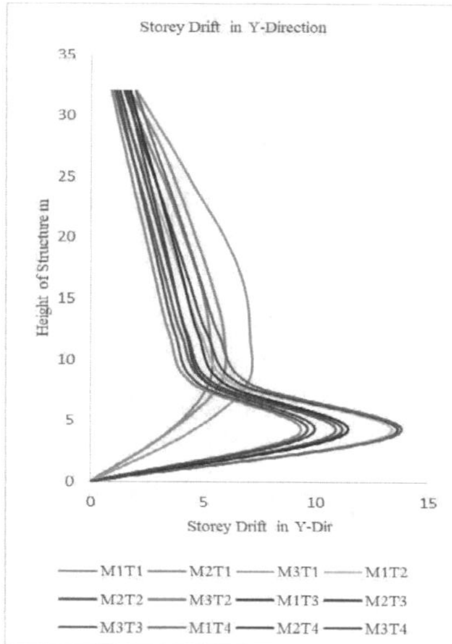

Figure 5. Comparison of story drift in the Y direction.

In figure 4, it has been observed that when the concrete grade increases the story drift decrease. The drift response for all the models in X-direction has been obtained at various story levels. The story drifts for concrete grade M20 has been decreased by 31.68%, 15.59% and 11.94% for M1T2, M1T3, and M1T4, respectively on the 8th story when compare to M1T1.

- For concrete grade M30 the value of story drift t has been decreased by 53.70%, 40.68% and 16.63% for M2T2, M2T3, and M2T4, respectively at 8th story when compare to M2T1.For concrete grade M40 the value of story drift has been decreased by 63.22%, 51.99% and 30.43% for M3T2, M3T3, and M3T4, respectively on the 8th story when compare toM3T1.
- The story drifts for concrete grade M20 has been decreased by 39.48%, 29.61% and 11.15% for M1T2, M1T3, and M1T4, respectively at 7th story when compare to M1T1. For concrete grade M30 the value of story drift has been decreased by 58.29%, 50.24% and 35.09% for M2T2, M2T3, and M2T4, respectively at 7th story when compare to M2T1. Forconcrete

grade M40 the value of story drift has been decreased by 67.87%, 61.17% and 47.97% for M3T2, M3T3, and M3T4, respectively at 7th story when compare to M3T1.

- The story drifts for concrete grade M20 has been decreased by 34.43%, 26.23% and 11.48% for M1T2, M1T3, and M1T4, respectively at 6th story when compare to M1T1. For concrete grade M30 the value of story drift has been decreased by 60.11%, 54.08% and 42.77% for M2T2, M2T3, and M2T4, respectively at 6th story when compare to M2T1. For concrete grade M40 the value of story drift has been decreased by 69.20%, 64.17% and 54.34% for M3T2, M3T3, and M3T4, respectively at 6th story when compare to M3T1.
- The story drifts for concrete grade M20 has been decreased by 34.70%, 28.18% and 16.32% for M1T2, M1T3, and M1T4, respectively at the 5th story when compare to M1T1.For concrete grade M30 the value of story drift has been decreased by 60.08%, 55.14% and 46% for M2T2, M2T3, and M2T4, respectively at 5th story when compare to M2T1. For concrete grade M40 the value of story drift has been decreased by 68.81%, 64.64% and 56.53% for M3T2, M3T3, and M3T4, respectively at 5th story when compare to M3T1.
- The story drifts for concrete grade M20 has been decreased by 29.89%, 24.30% and 13.34% for M1T2, M1T3, and M1T4, respectively at 4th story when compare to M1T1.
- For concrete grade M30 the value of story drift has been decreased by 58.20%, 53.96% and 46.20% for M2T2, M2T3, and M2T4, respectively at 4th story when compare to M2T1.For concrete grade M40 the value of story drift t has been decreased by 67.07%, 63.36% and 56.25% for M3T2, M3T3, and M3T4, respectively at 4th story when compare to M3T1.
- The story drift t for concrete grade M20 has been decreased by 25.18%, 19.72% and 9.97% for M1T2, M1T3, and M1T4, respectively at 3rd story when compare to M1T1.
- For concrete grade M30 the value of story drift has been decreased by 53.84%, 49.84% and 42.80% for M2T2, M2T3, and M2T4, respectively at 3rd story when compare to M2T1.For concrete grade M40 the value of story drift has been decreased by 64.52%, 61.04% and 54.71% for M3T2, M3T3, and M3T4, respectively at 3rd story when compare to M3T1.
- The story drifts for concrete grade M20 has been decreased by 20.44%, 12.43% and 4.71% for M1T2, M1T3, and M1T4, respectively at 2nd story when compare to M1T1.
- For concrete grade M30 the value of story drift has been decreased by 40.71%, 36.23% and 29.34% for M2T2, M2T3, and M2T4, respectively at 2nd story when compare to M2T1.For concrete grade M40 the value of story drift has been decreased by 54.44%, 50.99% and 44.065% for M3T2, M3T3, and M3T4, respectively at 2nd story when compare to M3T1.
- The story drifts for concrete grade M20 has been increased by 255.81%, 260.20% and 272.56% for M1T2, M1T3 and M1T4, respectively at 1st story when compare to M1T1.For concrete grade M30 the value of story drift has been increased by 104.74%, 111.78% and 118.72% for M2T2, M2T3, and M2T4, respectively at 1st story when compare to M2T1.For concrete grade M40 the value of story drift has been increased by 52.53%, 58.19% and 64.73% for M3T2, M3T3, and M3T4, respectively at 1st story when compare to M3T1.

In figure 5, it has been observed that when the concrete grade increases the story drift decrease. The drift response for all the models in X-direction has been obtained at various story levels. The story drifts for concrete grade M20 has been decreased by 41.91%, 23.93% and 19.68% for M1T2, M1T3, and M1T4, respectively on the 8th story when compare to M1T1.

- For concrete grade M30 the value of story drift t has been decreased by 38.66%, 24.58% and 13.84% for M2T2, M2T3, and M2T4, respectively at 8th story when compare to M2T1.For concrete grade M40 the value of story drift has been decreased by 54.32%, 42.45% and 19.91% for M3T2, M3T3, and M3T4, respectively at 8th story when compare to M3T1.
- The story drifts for concrete grade M20 has been decreased by 50.66%, 41.65% and 31.84% for M1T2, M1T3, and M1T4, respectively on the 7th story when compare to M1T1.For concrete grade M30 the value of story drift has been decreased by 45.56%,

37.10% and 21.32% for M2T2, M2T3, and M2T4, respectively at 7th story when compare to M2T1.For concrete grade M40 the value of story drift has been decreased by 58.54%, 51.25% and 37.39% for M3T2, M3T3, and M3T4, respectively at 7th story when compare to M3T1.

- The story drifts for concrete grade M20 has been decreased by 33.58%, 25.61% and 27.56% for M1T2, M1T3, and M1T4, respectively at 6th story when compare to M1T1.
- For concrete grade M30 the value of story drift has been decreased by 42.39%, 35.52% and 22.91% for M2T2, M2T3, and M2T4, respectively at 6th story when compare to M2T1.For concrete grade M40 the value of story drift has been decreased by 58.88%, 53.16% and 42.58% for M3T2, M3T3, and M3T4, respectively at 6th story when compare to M3T1.
- The story drifts for concrete grade M20 has been decreased by 22.59%, 15.45% and 56.13% for M1T2, M1T3, and M1T4, respectively at the 5th story when compare to M1T1.For concrete grade M30 the value of story drift has been decreased by 39.44%, 33.65% and 22.92% for M2T2, M2T3, and M2T4, respectively at 5th story when compare to M2T1.For concrete grade M40 the value of story drift has been decreased by 59.24%, 54.70 and 46.24% for M3T2, M3T3, and M3T4, respectively at 5th story when compare to M3T1.
- The story drifts for concrete grade M20 has been decreased by 21.52%, 16.19% and 21.90% for M1T2, M1T3, and M1T4, respectively on the 4th story when compare to M1T1.For concrete grade M30 the value of story drift has been decreased by 37.03%, 32.19% and 22.87% for M2T2, M2T3, and M2T4, respectively at 4th story when compare to M2T1.For concrete grade M40 the value of story drift t has been decreased by 57.01%, 53.125% and 45.87% for M3T2, M3T3, and M3T4, respectively at 4th story when compare to M3T1.
- The story drift t for concrete grade M20 has been decreased by 29.15%, 25.10% and 17.79% for M1T2, M1T3, and M1T4, respectively at 3rd story when compare to M1T1.
- For concrete grade M30 the value of story drift has been decreased by 31.86%, 27.37% and 19.02% for M2T2, M2T3, and M2T4, respectively at 3rd story when compare to M2T1.For concrete grade M40 the value of story drift has been decreased by 50.59%, 46.84% and 40.09% for M3T2, M3T3, and M3T4, respectively at 3rd story when compare to M3T1.
- The story drifts for concrete grade M20 has been decreased by 20.05%, 16.57% and 10.99% for M1T2, M1T3, and M1T4, respectively at 2nd story when compare to M1T1.
- For concrete grade M30 the value of story drift has been decreased by 8.811%, 33.91% and 4.27% for M2T2, M2T3, and M2T4, respectively at 2nd story when compare to M2T1.For concrete grade M40 the value of story drift has been decreased by 34.66%, 30.78% and 24.18% for M3T2, M3T3, and M3T4, respectively at 2nd story when compare to M3T1.
- The story drifts for concrete grade M20 has been increased by 292.63%, 300% and 304.61% for M1T2, M1T3 and M1T4, respectively at 1st story when compare to M1T1.
- For concrete grade M30 the value of story drift has been increased by 216.667%, 225% and 232.16% for M2T2, M2T3, and M2T4, respectively at 1st story when compare to M2T1.For concrete grade M40 the value of story drift has been increased by 112.67%, 119.284% and 126.26% for M3T2, M3T3, and M3T4, respectively at 1st story when compare to M3T1.

6.2. *Storey Displacement*

It has been observed that on the dynamic behaviour of the seismic response the displacement response increases from the ground floor to the upper floor in a building structure as shown in figure 6 and figure 7.

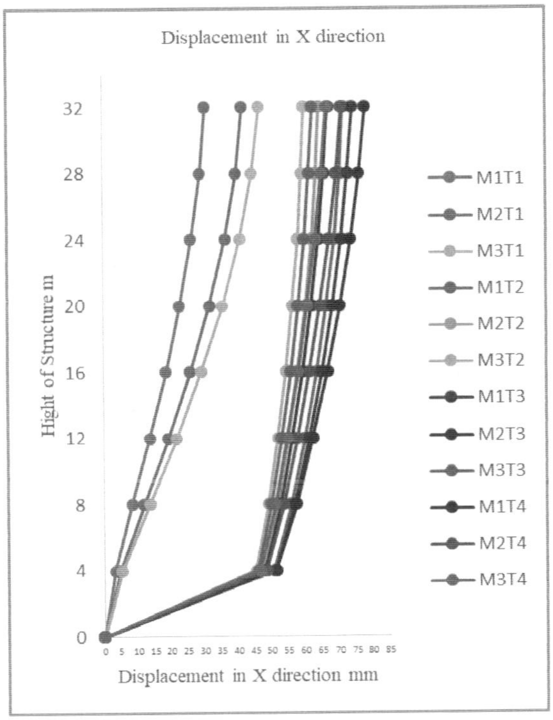

Figure 6. Comparison Story Displacement in the X direction.

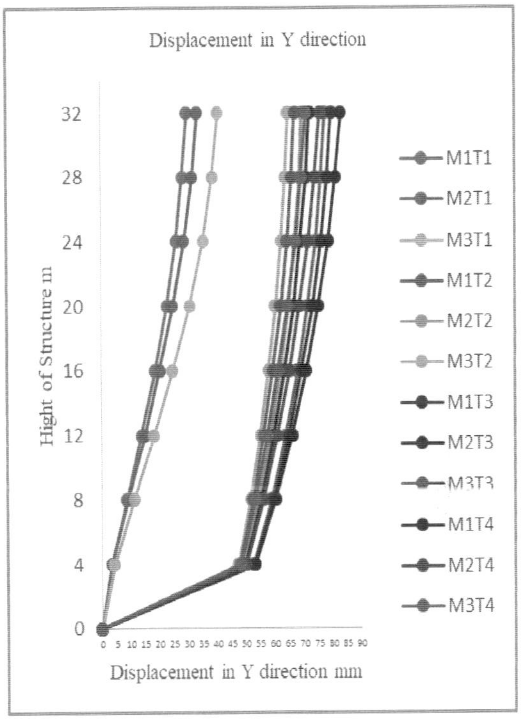

Figure 7. Comparison Story Displacement in Y direction.

In figure 6 it has been observed that when the concrete grade increases the story displacement also increases. The displacement response for all the models in X-direction has been obtained at various story levels. The story displacement for concrete grade M20 has been increased by 134.20%, 142.66% and 155.03% for M1T2, M1T3 and M1T4, respectively at the 8th story when compare to M1T1.

- For concrete grade M30 the value of story displacement has been increased by 54.32%, 60.66% and 70.37% for M2T2, M2T3, and M2T4, respectively at the 8th story when compare to M2T1.For concrete grade M40 the value of story displacement has been increased by 28.59%, 33.9% and 42.89% for M3T2, M3T3, and M3T4, respectively at 8th story when compare to M3T1.
- The story displacement for concrete grade M20 has been increased by 143.02%, 151.02% and 162.08% for M1T2, M1T3 and M1T4, respectively at the 7th story when compare to M1T1.For concrete grade M30 the value of story displacement has been increased by 59.52%, 65.66% and 74.51% for M2T2, M2T3, and M2T4, respectively at 7th story when compare to M2T1.For concrete grade M40 the value of story displacement has been increased by 32.89%, 37.9% and 46.29% for M3T2, M3T3, and M3T4, respectively at 7th story when compare to M3T1.
- The story displacement for concrete grade M20 has been increased by 162.23%, 170.02% and 180.50% for M1T2, M1T3 and M1T4, respectively at 6th story when

compare to M1T1.For concrete grade M30 the value of story displacement has been increased by 70.19%, 76.05% and 84.55% for M2T2, M2T3, and M2T4, respectively at 6th story when compare to M2T1.For concrete grade M40 the value of story displacement has been increased by 41.89%, 46.77% and 54.68% for M3T2, M3T3, and M3T4, respectively at 6th story when compare to M3T1.

- The story displacement for concrete grade M20 has been increased by 193.02%, 200.76% and 210.59% for M1T2, M1T3 and M1T4, respectively at 5th story when compare to M1T1.For concrete grade M30 the value of story displacement has been increased by 89.41%, 95.23% and 103.17% for M2T2, M2T3, and M2T4, respectively at 5th story when compare to M2T1.For concrete grade M40 the value of story displacement has been increased by 58.25%, 63.10% and 70.72% for M3T2, M3T3, and M3T4, respectively at 5th story when compare to M3T1.
- The story displacement for concrete grade M20 has been increased by 242.08%, 249.965% and 259.11% for M1T2, M1T3 and M1T4, respectively at 4th story when compare to M1T1.For concrete grade M30 the value of story displacement has been increased by 123.35%, 129.36% and 137.03% for M2T2, M2T3, and M2T4, respectively at 4th story when compare to M2T1.For concrete grade M40 the value of story displacement has been increased by 86.18%, 91.15% and 98.63% for M3T2, M3T3, and M3T4, respectively at 4th story when compare to M3T1.
- The story displacement for concrete grade M20 has been increased by 335.05%, 343.54% and 351.92% for M1T2, M1T3 and M1T4, respectively at 3rd story when compare to M1T1.For concrete grade M30 the value of story displacement has been increased by 186.35%, 192.94% and 200.03% for M2T2, M2T3, and M2T4, respectively at 3rd story when compare to M2T1.For concrete grade M40 the value of story displacement has been increased by 138.30%, 143.69% and 151.28% for M3T2, M3T3, and M3T4, respectively at 3rd story when compare to M3T1.
- The story displacement for concrete grade M20 has been increased by 565.79%, 576.18% and 583.57% for M1T2, M1T3 and M1T4, respectively at 2nd story when compare to M1T1.For concrete grade M30 the value of story displacement has been increased by 332.45%, 340.67% and 348.44% for M2T2, M2T3, and M2T4, respectively at 2nd story when compare to M2T1.For concrete grade M40 the value of story displacement has been increased by 258.83%, 265.37% and 273.78% for M3T2, M3T3, and M3T4, respectively at 2nd story when compare to M3T1.
- The story displacement for concrete grade M20 has been increased by 1403.1%, 1419.83% and 1423.57% for M1T2, M1T3 and M1T4, respectively at 1st story when compare to M1T1.For concrete grade M30 the value of story displacement has been increased by 903.19%, 917.70% and 926.78% for M2T2, M2T3, and M2T4, respectively at 1st story when compare to M2T1.For concrete grade M40 the value of story displacement has been increased by 732.29%, 743.68% and 755.67% for M3T2, M3T3, and M3T4, respectively at 1st story when compare to M3T1.

In figure 7 it has been observed that when the concrete grade increases the story displacement also increases. The displacement response for all the models in Y-direction has been obtained at various story levels. The story displacement for concrete grade M20 has been increased by 160.75%, 168.52% and 180.25% for M1T2, M1T3 and M1T4, respectively at 8th story when compare to M1T1.

- For concrete grade M30 the value of story displacement has been increased by 109.84%, 117.03% and 127.74% for M2T2, M2T3, and M2T4, respectively at 8th story when compare to M2T1.For concrete grade M40 the value of story displacement has been increased by 60.20%, 66.23% and 75.48% for M3T2, M3T3, and M3T4,respectively on the 8th story when compare to M3T1.
- j) The story displacement for concrete grade M20 has been increased by 170.09%, 177.36% and 187.88% for M1T2, M1T3 and M1T4, respectively at 7th storywhen

IOP Conf. Series: Materials Science and Engineering 1026 (2021) 012004 doi:10.1088/1757-899X/1026/1/012004

compare to M1T1.For concrete grade M30 the value of story displacement has been increased by 117.64%, 124.40% and 134.24% for M2T2, M2T3, and M2T4, respectively at 7th story when compare to M2T1.For concrete grade M40 the value of story displacement has been increased by 66.13%, 71.87% and 80.43% for M3T2, M3T3, and M3T4, respectively at 7th story when compare to M3T1.

- k) The story displacement for concrete grade M20 has been increased by 186.41%, 193.32% and 202.92% for M1T2, M1T3 and M1T4, respectively at 6th story when compare to M1T1.For concrete grade M30 the value of story displacement has been increased by 134.37%, 140.99% and 150.92% for M2T2, M2T3, and M2T4, respectively at 6th story when compare to M2T1.For concrete grade M40 the value of story displacement has been increased by 77.51%, 83.07% and 91.07% for M3T2, M3T3, and M3T4, respectively at 6th story when compare to M3T1.
- l) The story displacement for concrete grade M20 has been increased by 215.61%, 222.35% and 231.36% for M1T2, M1T3, and M1T4, respectively on the 5th story when compare to M1T1.For concrete grade M30 the value of story displacement has been increased by 161.82%, 168.39% and 176.98% for M2T2, M2T3, and M2T4, respectively at 5th story when compare to M2T1.For concrete grade M40 the value of story displacement has been increased by 97.96%, 103.50% and 111.10% for M3T2, M3T3, and M3T4, respectively at 5th story when compare to M3T1.
- m) The story displacement for concrete grade M20 has been increased by 266.61%, 273.40% and 280.78% for M1T2, M1T3 and M1T4, respectively at the 4th story when compare to M1T1.For concrete grade M30 the value of story displacement has been increased by 206.64%, 213.36% and 221.44% for M2T2, M2T3, and M2T4, respectively at 4th story when compare to M2T1.For concrete grade M40 the value of story displacement has been increased by 134.77%, 140.52% and 147.88% for M3T2, M3T3, and M3T4, respectively at 4th story when compare to M3T1.
- n) The story displacement for concrete grade M20 has been increased by 365.11%, 372.24% and 379.79% for M1T2, M1T3 and M1T4, respectively at 3rd story when compare to M1T1.For concrete grade M30 the value of story displacement has been increased by 292.04%, 299.34% and 306.94% for M2T2, M2T3, and M2T4, respectively at 3rd story when compare to M2T1.For concrete grade M40 the value of story displacement has been increased by 201.46%, 207.77% and 215.11% for M3T2, M3T3, and M3T4, respectively at 3rd story when compare to M3T1.
- o) The story displacement for concrete grade M20 has been increased by 585.3%, 593.49% and 600% for M1T2, M1T3 and M1T4, respectively at 2nd story when compare to M1T1.For concrete grade M30 the value of story displacement has been increased by 499.58%, 508.54% and 515.63% for M2T2, M2T3, and M2T4, respectively at 2nd story when compare to M2T1.For concrete grade M40 the value of story displacement has been increased by 358.28%, 366.18% and 373.65% for M3T2, M3T3, and M3T4, respectively at 2nd story when compare to M3T1.
- The story displacement for concrete grade M20 has been increased by 1455.07%, 1469.21% and 1472.46% for M1T2, M1T3 and M1T4, respectively at 1st story when compare to M1T1.For concrete grade M30 the value of story displacement has been increased by 1347.89%, 1363.67% and 1369.4% for M2T2, M2T3, and M2T4, respectively at 1st story when compare to M2T1.For concrete grade M40 the value of story displacement has been increased by 981.93%, 996.14% and 1005.03% for M3T2, M3T3, and M3T4, respectively at 1st story when compare to M3T1.

6.3. Spectral Acceleration.

It has been observed that on the dynamic behavior of the seismic response the spectral acceleration response increases as the grade of concrete increases as shown in figure 8.

- The spectral acceleration has been found to be maximum i.e. 0.756816g at response time 1.25 sec in the case of M40. Similarly the same has been found to be maximum i.e. 0.756816g at time 1.25 sec. The spectral acceleration in the case of M30 found to be 0.685512g at response time 1.265 sec. The spectral acceleration in the case of M20 found to be 0.393634g at response time 1.265sec. So, it can be inferred that as the grade of concrete increase the spectral acceleration also increases.

6.4. *Spectral Displacement*

It has been observed that on the seismic response, the spectral displacement response increases as the time response increases, it has been also observed that higher the concrete grade the higher will be the spectral displacement. Such spectral displacement response has been deployed in the forthcoming section as shown in figure 9.

- The value of spectral displacement has been increased for all the models (M1T1, M1T2 M1T3, and M1T4) of concrete grade M20. The spectral displacement has been found to be maximum i.e. 474.663mm at response time 5sec in the case of modal M1T4 for concrete M20. The spectral displacement has been found to be maximum i.e. 461.925mm at response time 5sec in the case of modal M1T3 for concrete M20. The spectral displacement has been found to be maximum i.e. 453.861mm at response time 5sec in the case of modal M1T3 for concrete M20. The spectral displacement has been found to be maximum i.e. 319.817mm at response time 5sec in the case of modal M1T1 for concrete M20.

6.5. *Spectral Velocity*

It has been observed that the dynamic behavior of the seismic response of the spectral velocity response increases as the grade of concrete increases. Such a spectral velocity response has been deployed in the forthcoming section as shown in figure 10.

- The spectral velocity has been found to be maximum i.e. 1391.4 mm/sec at response time 1.39 sec in the case of M40. Similarly the same has been found to be maximum i.e. 1391.4 mm/sec at response time 1.39 sec. The spectral velocity in the case of M30 found to be 1325.7 mm/sec at response time 1.265 sec. The spectral velocity in the case of M20 found to be 769.8mm/sec at response time 1.265sec. So, it can be inferred that as the grade of concrete increase the spectral velocity also increases.

6.6. *Base Shear*

Base shear is an estimate of the maximum expected lateral force that will occur at the base of a structure due to seismic ground motion. Infect base shear is a similar cumulative story shear from top to bottom story of the building structure as shown in figure 11.

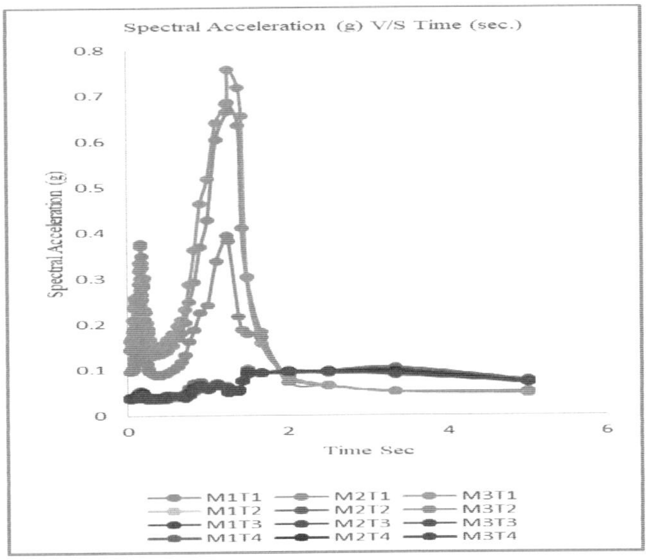

Figure 8. Comparison Spectral Acceleration in the X direction.

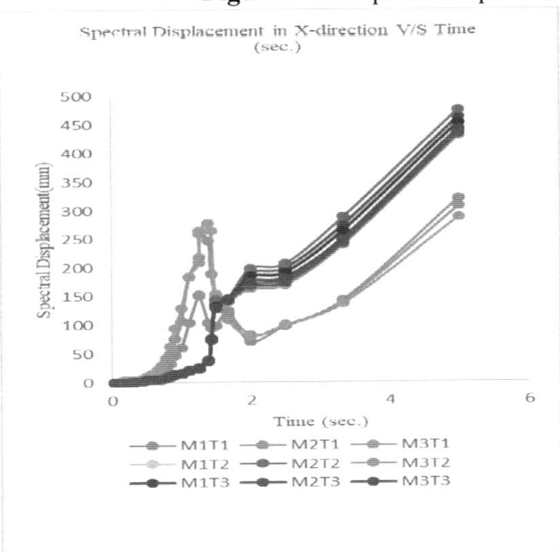

Figure9. Comparison Spectral displacement in X direction.

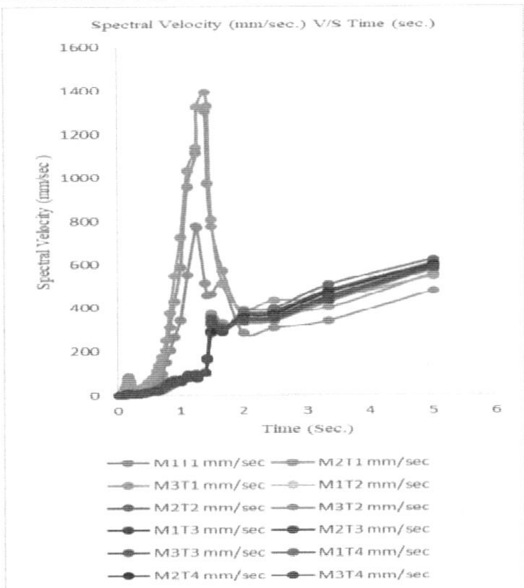

Figure 10. Comparison of Spectral velocity in the X-direction.

The base shear response for all the models in X-direction has been obtained. The base shear for concrete grade M20 has been decreased by150.10 %, 149.61% and 149.003% for M1T2, M1T3, and M1T4, respectively when compare to M1T1.For concrete grade M30 the value of base shear has been decreased by 147.14%, 147.14% and 146.038% for M2T2, M2T3, and M2T4, respectively when compared to M2T1.For concrete grade M40 the value of base shear has been decreased by 145.024%, 144.56% and 143.89% for M3T2, M3T3, and M3T4, respectively when compared to M3T1.

6.7. *Lateral Shear*

It has been observed that the dynamic behaviour of the seismic response of the lateral shear forces response increases from the ground floor to the upper floor in a building structure. Such lateral shear forces response has been deployed in the fourth coming section as shown in figure 12.

- The lateral story force for all the models in X-direction has been obtained at various story levels. The lateral story force for concrete grade M20 has decreased by 49.89%, 50.38% and 50.98% for M1T2, M1T3, and M1T4, respectively at 8th story when compare to M1T1.

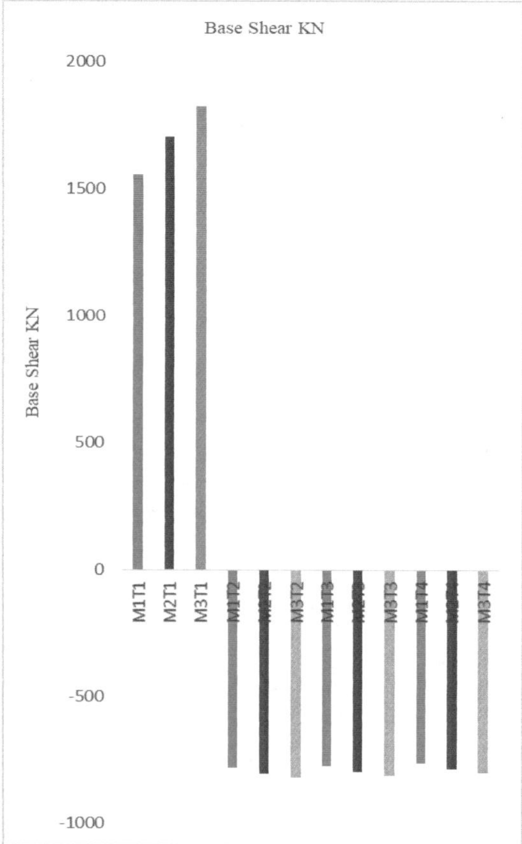

Figure 11. Comparison Base shear in X direction.

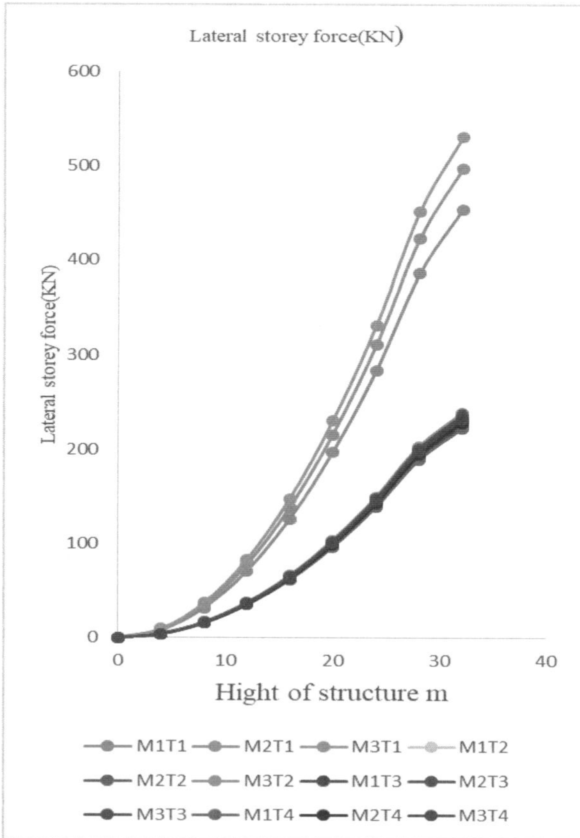

Figure 12. Comparison Lateral story force in the X-direction.

- For concrete grade M30 the value of lateral story force has decreased by 52.85%, 53.32% and 53.95% for M2T2, M2T3, and M2T4, respectively at the 8th story when compare to M2T1. For concrete grade M40 the value of lateral story force has decreased by 54.97%, 55.43% and 56.06% for M3T2, M3T3, and M3T4, respectively on the 8th story when compare to M3T1.
- The lateral story force for concrete grade M20 has decreased by 49.89%, 50.38% and 50.98% for M1T2, M1T3, and M1T4, respectively at 7th story when compare to M1T1.For concrete grade M30 the value of lateral story force has been decreased by 52.85%, 53.32% and 53.95% for M2T2, M2T3, and M2T4, respectively at 7th story when compare to M2T1.For concrete grade M40 the value of lateral story force has been decreased by 54.97%, 55.43% and 56.06%for M3T2, M3T3, and M3T4, respectively at 7th story when compare to M3T1.
- The lateral story force for concrete grade M20 has been decreased by 49.89%, 50.38% and 50.98% for M1T2, M1T3, and M1T4, respectively at 6th story when compare to M1T1.For concrete grade M30 the value of lateral story force has been decreased by 52.85%, 53.32% and 53.95% for M2T2, M2T3, and M2T4, respectively at 6th story when compare to M2T1.For concrete grade M40 the value of lateral story force has been decreased by54.97%,

IOP Conf. Series: Materials Science and Engineering 1026 (2021) 012004 doi:10.1088/1757-899X/1026/1/012004

55.43% and 56.06% for M3T2, M3T3, and M3T4, respectively at 6th story when compare to M3T1.

- The lateral story force for concrete grade M20 has been decreased by 49.89%, 50.38% and 50.98% for M1T2, M1T3, and M1T4, respectively at 5th story when compare to M1T1.For concrete grade M30 the value of lateral story force has been decreased by 52.85%, 53.32% and 53.95% for M2T2, M2T3, and M2T4, respectively at 5th story when compare to M2T1.For concrete grade M40 the value of lateral story force has been decreased by 54.97%, 55.43% and 56.06% for M3T2, M3T3, and M3T4, respectively at 5th story when compare to M3T1.
- The lateral story force for concrete grade M20 has been decreased by 49.89%, 50.38% and 50.98% for M1T2, M1T3, and M1T4, respectively at 4th story when compare to M1T1.For concrete grade M30 the value of lateral story force has been decreased by 52.85%, 53.32% and 53.95%for M2T2, M2T3, and M2T4, respectively at 4th story when compare to M2T1.For concrete grade M40 the value of lateral story force has been decreased by 54.97%, 55.43% and 56.06% for M3T2, M3T3, and M3T4, respectively at 4th story when compare to M3T1.
- The lateral story force for concrete grade M20 has been decreased by 49.89%, 50.38% and 50.98% for M1T2, M1T3, and M1T4, respectively at 3rd story when compare to M1T1.For concrete grade M30 the value of lateral story force has been decreased by 52.85%, 53.32% and 53.95% for M2T2, M2T3, and M2T4, respectively at 3rd story when compare to M2T1.For concrete grade M40 the value of lateral story force has been decreased by 54.97%, 55.43% and 56.06% for M3T2, M3T3, and M3T4, respectively at 3rd story when compare to M3T1.
- The lateral story force for concrete grade M20 has been decreased by 49.89%, 50.38% and 50.98% for M1T2, M1T3, and M1T4, respectively at 2nd story when compare to M1T1.For concrete grade M30 the value of lateral story force has been decreased by 52.85%, 53.32% and 53.95% for M2T2, M2T3, and M2T4, respectively at 2nd story when compare to M2T1.For concrete grade M40 the value of lateral story force has been decreased by 54.97%, 55.43% and 56.06% for M3T2, M3T3, and M3T4, respectively at 2nd story when compare to M3T1.
- The lateral story force for concrete grade M20 has been decreased by 50.02%, 50.63% and 51.35% for M1T2, M1T3, and M1T4, respectively at 1st story when compare to M1T1.For concrete grade M30 the value of lateral story force has been decreased by 52.97%, 53.56% and 54.30% for M2T2, M2T3, and M2T4, respectively at 1st story when compare to M2T1.For concrete grade M40 the value of lateral story force has been decreased by 55.08%, 55.65% and 56.40% for M3T2, M3T3, and M3T4, respectively at 1st story when compare to M3T1.

6.8. Frequency

It has been observed that the dynamic behavior of the seismic response of the frequency increases as modes increases in a building structure. Such a frequency response has been deployed in the forthcoming section as shown in figure 13.

- The frequency has been found to be maximum i.e. 13.333Hz/sec at response mode 20 in the case grade of concrete increase, the frequency of the modal system also increases. It has been found that the 13.333Hz/sec in the case of M40 and the same for M30, M20 respectively. The frequency in the case of M30 found to be 11.23596Hz/sec at response mode 20. The frequency in the case of M20 found to be 12.596Hz/sec at response mode 20. So, it can be inferred that as the grade of concrete increase the spectral frequency also increases.

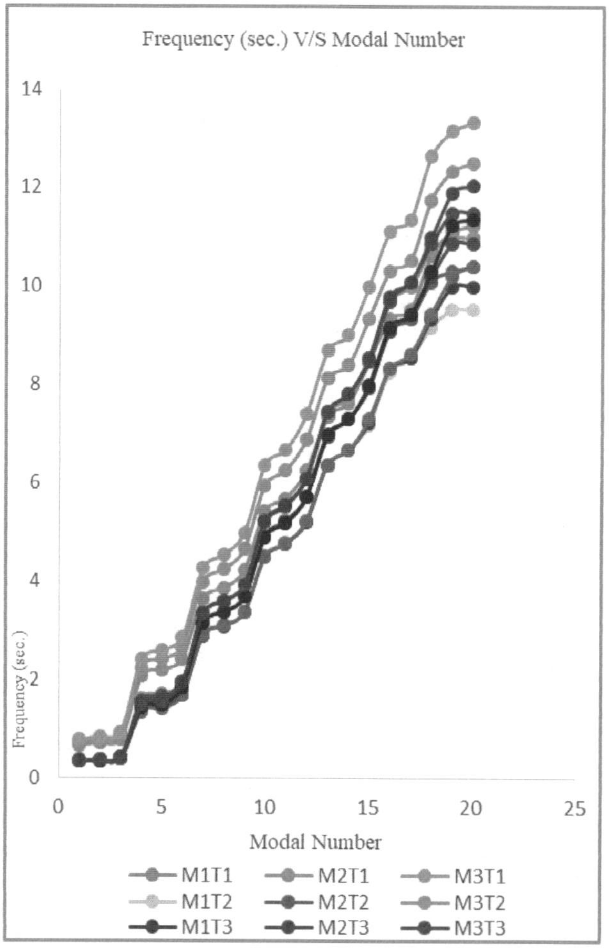

Figure 13. Comparison Frequency in X direction.

7. Conclusion

The seismic analysis of fixed base and base-isolated RCC frames has been carried out. 12 types M1T1, M2T1, M3T1, M1T2, M2T2, M3T2, M1T3, M2T3, M3T3, M1T4, M2T4, and M3T4 have been used under seismic zone IV conditions. The following are the conclusion of the study.

- The flexibility at the base of the building increases with the use of rubber bearing isolators.
- The use of base isolation decreases the story drift, increases displacement and gives better seismic protection during earthquakes...
- In 100mm, 200mm, and 300mm thickness of lead rubber isolated building models for concrete grade M20, story drift is reduced by 31.68%, 15.59% and 11.94% in the X direction and 41.91%, 23.93% and 19.68% in Y direction compared to the fixed base model.
- In 100mm, 200mm, and 300mm thickness of lead rubber isolated building models for concrete grade M30, story drift is reduced by 53.70%, 40.68% and 16.63% in the X direction and 38.66%, 24.58% and 13.84% in Y direction compared to the fixed base model.
- In 100mm, 200mm, and 300mm thickness of lead rubber isolated building models for concrete grade M40, story drift is reduced by 63.22%, 51.99% and 30.43% in the X direction and 54.32%, 42.45% and 19.91% in Y direction compared to the fixed base model.
- In 100mm, 200mm, and 300mm thickness of lead rubber isolated building models for concrete grade M20, story displacement is increased by 134.20%, 142.66% and 155.03% in

IOP Conf. Series: Materials Science and Engineering 1026 (2021) 012004 doi:10.1088/1757-899X/1026/1/012004

the X direction and 160.75%, 168.52% and 180.25% in Y direction compared to the fixed base model.

- In 100mm, 200mm, and 300mm thickness of lead rubber isolated building models for concrete grade M30, story displacement is increased by 54.32%, 60.66% and 70.37% in the X direction and 109.84%, 117.03% and 127.74%in Y direction compared to the fixed base model.
- In 100mm, 200mm, and 300mm thickness of lead rubber isolated building models for concrete grade M40, story displacement is increased by 28.59%, 33.9% and 42.89% in X direction and 60.20%, 66.23% and 75.48% in Y direction compared to the fixed basemodel.
- In spectral acceleration as the grade of concrete increases, the acceleration peak also increases as been found to be in the case of M40, M30, with respect to M20.
- In spectral displacement it has been found that the value of spectral displacement has been increased for all the models with lead rubber base isolator (M1T2, M1T3, and M1T4) of concrete grade M20. And in spectral displacement as the grade of concrete increases, the displacement peak decreases. And it has been found to be in case of M40, M30, with respect to M20 in case of lead rubber isolated base of the story.
- In spectral velocity as the grade of concrete increased the velocity peak also increased. And it has been found to be in case of M40, M30, with respect to M20 in case of fixed and isolated base of the story.
- Base shear was decreased when the lead rubber base isolator being operational. The percentage of reduction has been found i.e. 150.10 %, 149.61% and 149.003% in X direction to be in case of M20 with thickness of base isolator.100mm, 200mm, and 300mm.
- Base shear was decreased when the lead rubber base isolator being operational. The percentage of reduction has been found i.e. 147.14%, 147.14% and 146.038% in X direction to be in case of M30 with thickness of base isolator.100mm, 200mm, and 300mm.
- Base shear was decreased when the lead rubber base isolator being operational. The percentage of reduction has been found i.e. 145.024%, 144.56% and 143.89% % in X direction to be in case of M40 with thickness of base isolator.100mm, 200mm, and 300mm.
- Lateral story forces decreased when the lead rubber base isolator being operational. The percentage reduction has been found i.e. 49.89%, 50.38% and 50.98% in X direction to be in case of M20 with thickness of base isolator.100mm, 200mm, and 300mm.
- Lateral story forces decreased when the lead rubber base isolator being operational. The percentage reduction has been found i.e. 52.85%, 53.32% and 53.95% in the X direction to be in case of M20 with a thickness of base isolator.100mm, 200mm, and 300mm.
- Lateral story forces decreased when the lead rubber base isolator being operational. The percentage reduction has been found i.e. 54.97%, 55.43% and 56.06% in the X direction to be in case of M20 with a thickness of base isolator.100mm, 200mm, and 300mm.
- The frequency increases as the strength of the concrete increases. However, the variation increases in the frequency is to the extent for concrete grade M20 found to be 11.596Hz/sec at response on mode 20. For concrete grade M30 found to be 12.23596Hz/sec at the response on mode 20. For concrete grade M40 found to be 13.3333Hz/sec at the response on mode 20.
- The effect of higher energy seismic waves mitigated by using the different thickness of lead rubber base isolators
- It is more important to isolate the structure from the source of seismic force throughsome medium of the material of stiffness lesser than the stiffness of the structure.

8. Recommendation

- More study on base isolation needs to be done in view of the various base isolators available in the market.
- More study can be done on 100mm 200mm 300mm lead rubber base isolators withmore concrete grade.
- More study can be done on the geometry of base isolators.

Acknowledgment
I would like to express my deepest appreciation to all those who provided me the possibility to complete this study. A special gratitude I give to Shoolini University of Biotechnology and Management, whose contribution in stimulating suggestions and encouragement helped me to coordinate my work especially in writing this study.
I express my sincere thanks to SRM institute of science and technology for support and timely help.

References

[1] A N Lin. Seismic Performance of Fixed-Base and Base-Isolated Steel Frames. *J. Eng. Mech. 1992*; Vol.**118**:921-941.
[2] Anil.K.Chopra (2007), „Text book on Dynamics of Structures -Theory and Application to Earthquake Engineering", Prentice Hall International series.
[3] A. Ras et al. Dissipation Capacity of Steel Building with Fiction Pendulum Base-Isolation System. World Academy of Science, Engineering and Technology *International Journal of Computer and Systems Engineering*. 2016; Vol.**10**, No:11.
[4] Azlan ADNAN et al. Seismic Base Isolation of Steel Frame Structure by Hollow Rubber Bearings. *Gazi University Journal of Science*. 2011; Vol.**24 (4)**:841-853.
[5] Bungale S. Taranath, Steel, Concrete, and Composite Design of Tall Building . CRC Press Taylor and Francis Group.
[6] Cancellara Donato and Angeli Fabio De. Nonlinear dynamic analysis for multi-storey RC structures with hybrid base isolation systems in presence of bi-directional ground motions. *Composite structure*. **2016**.
[7] Chang Chia-Ming et al. Design of Buildings with Seismic Isolation Using Linear Quadratic Algorithm. *Procedia Engineering* **199 (2017)** 1610–1615.
[8] ETABS, Version (16.2.1), *Release Notes* **16.2.1**, Inc., 2016.
[9] Fujita Takafumi. Seismic isolation rubber bearings for nuclear facilities Nuclear Engineering and Design 127 (1991) 379-391
[10] Hadian A. V, Mutalib A. and BaharomA S. Seismic Behaviour Of Base Isolation System Using Lead Rubber Bearing. 2013; *ISSN* **0127**–9696.
[11] IS: 875 Part 1-5 (1987)„„Code of practice for Design Load (other than earthquake) for Buildings and Structures" , Bureau of Indian Standard, New Delhi.
[12] IS: 1893 Part I (2002), „Code of practice for Criteria for Earthquake Resistance Design of Structure :General provisions and Building" , Bureau of Indian Standard, NewDelhi.
[13] IS: 4326 (1993)„„Code of practice for Earthquake Resistant Design and Construction of Buildings (Second Revision)" , Bureau of Indian Standard, New Delhi.
[14] IS: 456 (2000)„„Code of practice for Plan and Reinforced Concrete (fourth revision)" , Bureau of Indian Standard, New Delhi.
[15] IS: 13920 (1993)„„Code of practice for Ductile detailing of Reinforced Concrete Structures Subjected to Seismic Forces" , Bureau of Indian Standard, New Delhi.
[16] Kilar Vojko and Koren David. Seismic behaviour of asymmetric base isolated structures with various distributions of isolators. *Engineering Structures* **31 (2009)** 910_921.
[17] Kiran K Shetty, Krishnamoorthy et al. Dynamic Response Control of Base Isolated Plane Frame Structure Using Multiple Tuned Mass Dampers, *Bonfring International Journal of Industrial Engineering and Management Science*. 2012; Vol. **2**, No. 4.
[18] M.Vijayakumar et al. A Study on Seismic Performance of RCC Frame with Various Bracing Systems using Base Isolation Technique. *International Journal of Applied Engineering Research*. 2016; Vol. **11**:7030-7033.
[19] Minal Ashok Somwanshi and Mrs. Rina N. Pantawane. Seismic Analysis of Fixed Based and Base Isolated Building Structures. *International Journal of Multidisciplinary and Current Research*. 2015; Vol.**3**:747-757.
[20] Mital N. Desai and Prof. Roshni John. Comparative Study of Multi-Storey Building

with Different Base Isolators. *International Journal of Innovative Research in Science, Engineering and Technology*. 2015; Vol. **4**:7353-7360.

[21] Moon Byung-Young and Kang Gyung-Ju. Design and manufacturing of fiber reinforced elastomeric isolator for seismic isolation. *Journal of Materials Processing Technology* **130–131 (2002)** 145–150.

[22] Paz Mario. Structural Dynamics Theory and Computaion. Fourth edition. 1997.

[23] Piero Colajanni and Maurizio Papia. Seismic response of braced frames with and without friction dampers, *Engineering Structure*. 1995; Vol. **17** Number 2.

[24] Rathi Prof. V. R. and Gadakh Vishal Balkrishna. Effect of Base Isolation on Building. *International Journal of Modern Trends in Engineering and Research (IJMTER) 2016*; Vol. **3**:581-589.

[25] Ryota Nakamura and Minoru Yamanari. Control of Vibration in Steel Structures by Base-isolation System Using Friction Dampers, *J. Temporal Des. Arch. Environ. 2013*; VOL.**12** (1):144-148.

[26] Satish Nagarajaiah. Base-Isolated FCC Building:Impact Response in Northridge Earthquake. *Journal of Structural Engineering*. **2001**; 1063-1075.

[27] Y. Zhou, S.M. Lin, C.X Wu and X.S. Deng "Analysis of high – rise building with energy dissipation storey system " 15 WCEE Lisboa 2012.

[28] Stefano Sorace et al. Experimental investigation on a base isolation system in corporating steel–Teflon slider sand pressurized fluid viscous spring dampers. *Earthquake Engineering and Structural Dynamics Earthquake Engng Struct. Dyn. 2008*; Vol.**37**:225–242.

[29] Venkatesh M. B. and R. D. Deshpande. Analysis of R.C. building frame with raft foundation considering soil structure interaction, 2017; **752-760**.

[30] Viswanath K.G, Seismic. Analysis of Steel Braced Reinforced Concrete Frames. *International Journal of Civil and Structural Engineering. 2010*; Vol. **1**:114-122.

[31] Y. L. MO and Y. F. Chang. Application of Base Isolation Concept to Soft First Story Buildings. *Compureu & Svurruws. 1995*; Vol. **55**:883-896.

An approach to derive seismic fragility curves

V J Vedhanayaghi[1*], M Manoharan[1], J Jasper Daniel[1], S Premkumar[1], S ArunBharathi[2]

[1]Faculty, Department of Civil Engineering, Rajalakshmi Engineering College, Chennai – 602 105

[2]Faculty, Department of Civil Engineering, Misrimal Navajee Munoth Jain Engineering College, Chennai – 600 097

*Corresponding author email:vedhanayaghi.vj@gmail.com

Abstract Seismic risk assessment is needed for the evaluation of disaster management, emergency preparedness and prior retrofitting. Based on feasible collapse mechanisms and calculation of their associated seismic parameter Peak Ground Displacement, the failure analysis of the structure is identified. Further, the outputs are manipulated to produce a measure vulnerability obtained from failure analysis. In regions of moderate and high seismicity, Zone – III Fragility analysis is used. An investigation is made on low rise Unreinforced masonry (URM) structure located in Zone – III, particularly in Chennai. To obtain capacity curve and bi-linear capacity curve the structure is modelled as Equivalent Frame Model (EFM), and non – linear static pushover analysis (NPOA).The structure damages, under various level of seismic excitation was developed by an analytical seismic fragility curve. For prioritizing retrofit, pre-earthquake planning, and loss estimation tools such as cost-benefit analysis this curve is utilized. The damage states considered for determining probability of exceeding a specified damage state involved in fragility curve generation are slight, moderate, extreme and collapse.
Key words: *vulnerability, retrofitting, PGD, fragility curves, seismic assessment, URM.*

1. Introduction

Those structures which are built from individual units laid in and bound together by mortar are said to be masonry structures. Without the intervention of skilled labours till twentieth century masonry structures were constructed in Chennai. Poor performance in masonry structures results in failure of structure during earthquake masonry structures are well known for economical construction and simplicity.
Fragility curve refers to the probability of reaching or exceeding a given damage state defined as a function of earthquake loading. Alternative forms of construction such as confined masonry or reinforced masonry, considered less vulnerable, have been developed. Empirical, judgmental, analytical and hybrid approaches are used to obtain fragility curve [11].

1.1. Need of the study

The concepts of evaluating seismic safety of URM structures are the primary part of this Project. The research on seismic behaviour of URM structures and developing fragility curves for URM buildings, subject to seismic loading are included in this study. The fragility curves can be directly used to perform a thorough risk analysis for historical structures.

1.2. Objective

The objective of this investigation is:

- To analyse current strength characteristics of URM structure.
- To describe different damage or limit state by varying seismic parameter.
- To introduce a procedure for developing fragility curves for typical URM structure in Chennai and to present the generated results.

1.3 Scope

- To determine strength characteristics, soil conditions and influence of age on seismic response by nonlinear static pushover analysis.
- To evaluate seismic performance of the structure by displacement coefficient method.
- To develop fragility curve by lognormal distribution method for historic masonry structures in Chennai.

2. Validation

Research papers by Ruiz Garcia, et.al., (2010) [17] and S.L.Dimova and K.Hirata (2000) [18] were validated.
SAP software was used for the Equivalent-frame modelling and SAP2000 for Non-linear static pushover analysis. The comparison with the results of the journal was done by generating fragility curves and adopting HAZUS [20] method. The same method applied on many other journals showed similar results up to 98.9%, hence this method will be effective.

3. Derivation process

The derivation process of analytical fragility curves for a particular building or building typology comprises several components [2]. They are as follows:

- Building Inventory,
- Damage/limit state model,
- Building capacity model,
- Seismic hazard analysis,
- Seismic demand model, and
- Fragility generation.

The methodology worked out to develop analytical fragility curves is shown in figure 1.

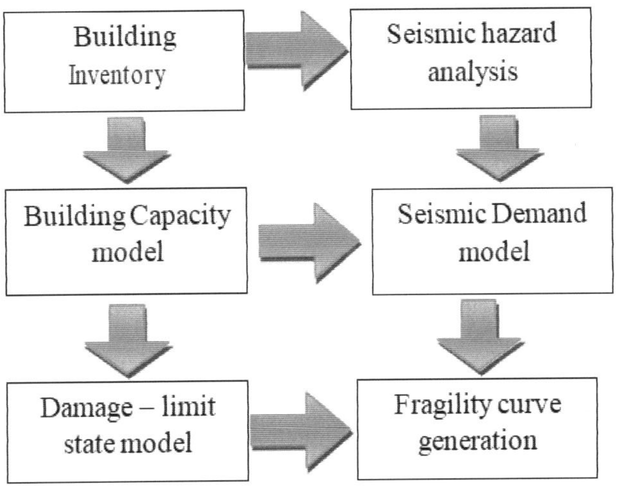

Figure 1. Derivation process – seismic fragility curves

IOP Conf. Series: Materials Science and Engineering 1026 (2021) 012007 doi:10.1088/1757-899X/1026/1/012007

4. Building inventory

Residential building in Old Washermenpet has been chosen and was constructed during the year 1920 and has successfully served its purpose for almost 93 long years. The building is a masonry load bearing type structure. Brick masonry was plastered with clay and egg albumen mortar. Renovation work was carried out during the year 1962 and 2000. The building is double storied, 8.8 m long and 9.2 m wide building. The foundation is stepped spread footing type, constructed along the length of walls at ground floor resting on Type –II soil as per IS 1893:2012. The ground floor and first floor are nearly 2.74 m tall. The structure chosen for analysis & it plan details is represented in figure 2.

(a) URM building

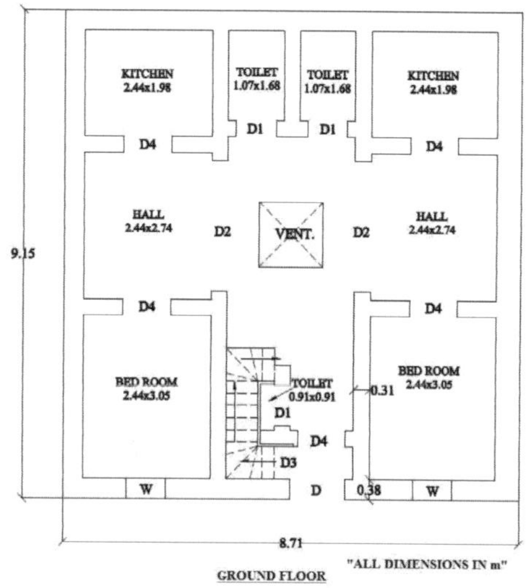

(b) Ground Floor Plan

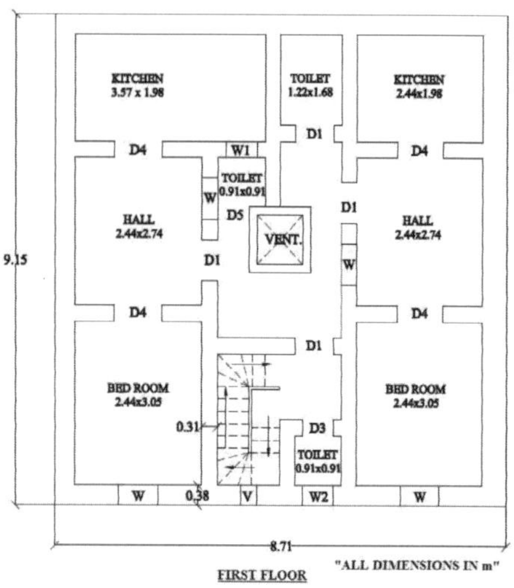

(c) First Floor Plan

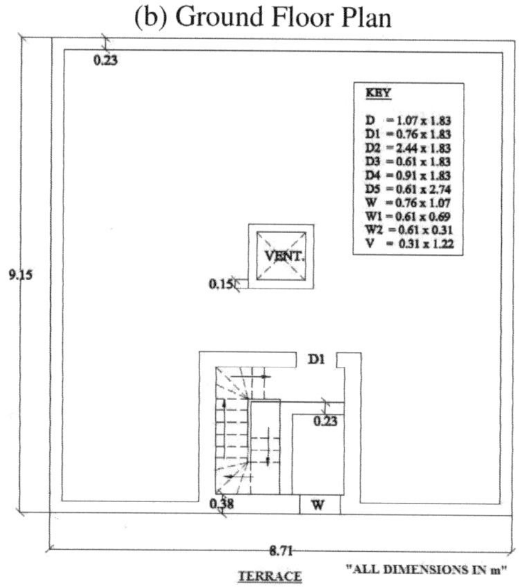

(d) Terrace Plan

Figure 2. Building under study

5. Seismic hazard analysis

In this investigation we focus on the seismic hazard analysis of Chennai, where the structure is located with respect to the seismic hazard map of India. It could be inferred that there is a chance of occurrence of moderate earthquake to occur in Chennai. Also, a study on the damages occurred to the structure due to past earthquakes is done [1, 3, 4, 10]. The damages occurred to the structure chosen due to past earthquakes is represented in table 1. The hairline cracks formed on the exterior walls of the building chosen for study due to the earthquake occurred during December 2004 is shown in figure 3.

Table 1. Earthquakes that affected Chennai in past 20 years (1994 – 2014).

Month & year of occurrence	Epicentre	Magnitude	Damage occurred to the building chosen
September 2001	Chennai	<3	Slight tremor was observed by persons living in the building
December 2004	250km SSE of Banda Aceh, Sumatra	9.1	Formation of hairline cracks on exterior wall
April 2012	Indian Ocean	8.6 and 8.2	Propagation of hairline crack formed during 2004 earthquake

Figure 3. Hairline cracks formed on exterior wall (arrow mark)

6. Modelling of the structure

The structure is modelled as an Equivalent Frame Model (EFM) to estimate the building capacity and seismic demand of the structure. The building capacity is determined from a plot of base shear versus roof displacement, whereas seismic demand is determined from a plot of spectral acceleration versus spectral displacement; both the plots are obtained from Non-linear static pushover analysis (NPOA) [5,6]. The mechanical properties of masonry units used during early 20th century were considered while modelling the structure. The properties are tabulated in table 2.

Table 2. Properties of masonry units considered for analysis

Properties considered	Values used
Weight per unit volume	18.4 kN/m^3
Mass per unit volume	1.8763 kN/m^3
Modulus of Elasticity (E)	1.6 x $10^6 kN/m^2$
Poisson's Ratio (μ)	0.23
Shear Modulus (G)	650406.5 kN/m^2
Co-efficient of thermal expansion	5.5 x 10^{-6}
Material geometry	• Anisotropic for Earthquake loading condition. • Orthotropic for Normal loading condition.

EFM of masonry walls has been used successfully in many previous studies for the assessment of seismic behaviour of masonry buildings in Bhuj, India. In this method, multi-storey masonry walls are modelled as equivalent frames made of vertical (pier) and horizontal (spandrels) elements with rigid intersecting joint elements. The equivalent frame model adopted in this study is shown in figure 4(a) for a typical masonry wall. The effective height of the pier is calculated according to the criteria of Dolce [1989] [7] and can be expressed as:

$$H_{eff} = h' + \left\{\left[\frac{1}{3h'} \times D \times (H - h')\right]\right\} \tag{1}$$

where, h', D, H for the given pier can be obtained as shown in figure 4 (b). The nonlinear behaviour of piers and spandrels is modelled by inserting plastic hinges at pre-defined locations in the frame elements, to overcome the limitation of in-ability of EFM to automatically simulate the effect of varying axial stresses on piers. The effective height of piers on left (L) and right (R) side of the opening in facade wall found as per formula 1 is tabulated in table 3.

Table 3. Effective height of piers as per Dolce expression

Notation of Opening	Height of Opening (m)	Height of Wall (m)	Pier Dimensions (m)				Effective Height of Pier, H_{eff} (m)	
			h'		D			
			L	R	L	R	L	R
D1	1.83	2.743	2.41	2.23	2.01	1.39	2.50	2.34
W1	1.07	2.743	1.47	1.45	1.39	1.32	1.87	1.84
W2	1.07	2.743	1.45	1.47	1.32	1.39	1.84	1.87
W3	1.07	2.743	1.47	1.65	1.39	2.01	1.87	2.09
W4	1.07	2.743	1.47	1.47	1.39	1.39	1.87	1.87
W	0.31	2.743	0.57	0.69	0.91	1.32	1.72	1.99
V	1.22	2.743	1.60	1.48	1.32	0.91	1.91	1.74

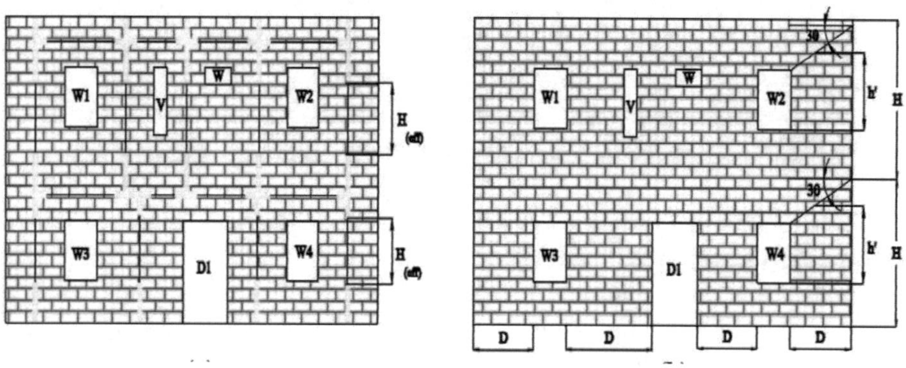

Figure 4. (a) Equivalent frame modelling of a typical multi-story masonry wall with door and window openings, and (b) Definition of effective height of piers as per Dolce [1989]

6.1. Nonlinear static pushover analysis

In this investigation we focus on the nonlinear static pushover analysis(NPOA) procedure in the FEMA – 273 and ATC 40. The floor displacements, story drifts, joint rotations, plastic hinge rotations, etc., computed at the target displacement represent the earthquake induced demands on the structure [8, 12].

The bilinear capacity curve, a plot of spectral acceleration versus spectral displacement is determined by NPOA of structure subject to spectral acceleration gradually increasing until a target value of spectral displacement is reached as per ATC 40 procedure. The capacity curve and bilinear capacity curve of the building analyzed are obtained and represented in figure 6 and 7. The EFM for the building analyzed and the deformed shape after performing NPOA is represented in figure 5.

The steps involved in carrying out NPOA on the EFM of structure studied using SAP2000 is given below:

- Set default unit as 'kN, m, C'
- Create grid lines.
- Define material specification – i.e. material properties.

- Define dimensions of lintel and frame
- Define supports
- Assign lintel and piers according to equivalent frame model procedure.
- Design Lintels and piers as RCC frame.
- Define DL & LL
- Define Response Spectrum as per IS 1893:2012 (this codal provision is included in SAP2000).
    - Select Zone-III and Soil type as Type-II (medium soil).
    - I = 1.0
    - R = 1.5 (for URM).
    - Sa/g = 2.50 for (T = 0.17s).
    - Z = 0.16.
    - Damping ratio = 5%
- Define Load case → DL + LL + EL
- Define nodes
- Define hinges – assign hinge using auto hinge property
- Assign loads.
- Define pushover load case with unit displacement along XX direction at node no 48 (node present at top extreme end)
- Run Analysis
- Display result

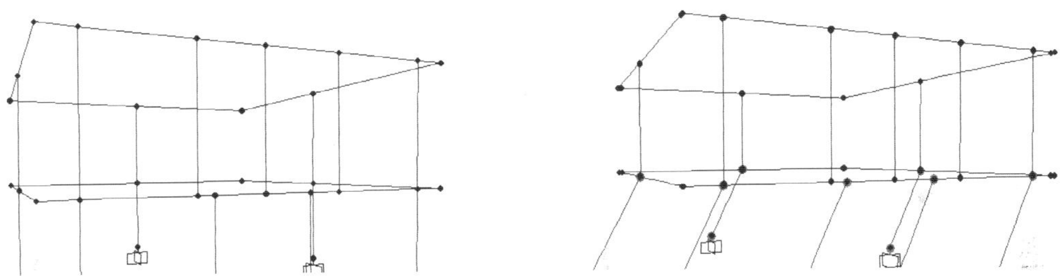

Actual model Deformed shape after performing NPOA

Figure 5. Building modelling

Pushover analysis of EFM of the structure studied resulted in generation of capacity curve for the structure as per FEMA – 273, shown in figure 6. The structure experienced a yield base shear of 2650.389kN and yield displacement of 45mm (0.045m), after which the structure collapses. The results of building capacity model are presented in table 4.

Table 4. Analysis result of building capacity model

Base Shear (kN)	Roof Displacement (m)
0	0
2335.1542	8.010×10^{-4}
2434.5762	1.602×10^{-3}
2645.3220	0.0320
2650.3899	0.045

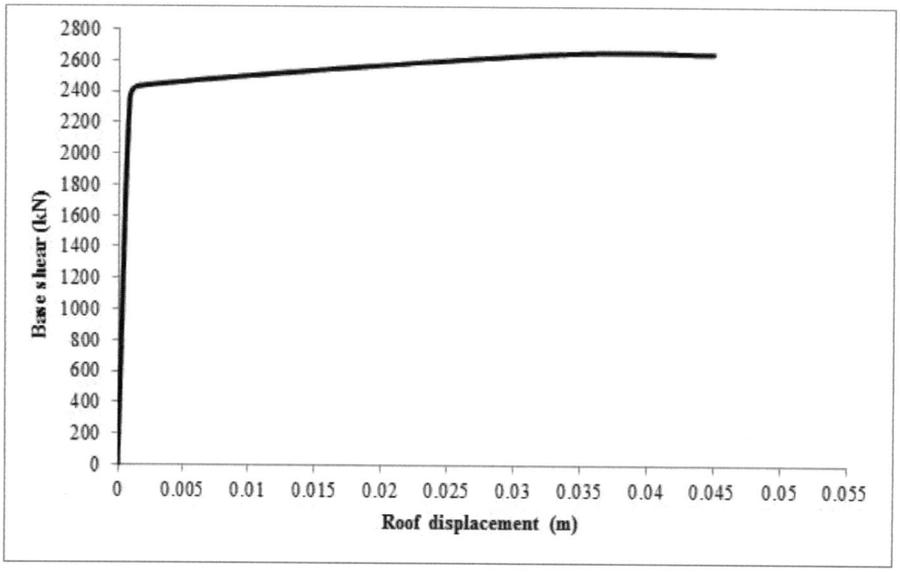

Figure 6. Capacity curve

NPOA of EFM of the structure studied resulted in generation of capacity curve for the structure as per ATC 40, bilinear capacity curve shown in figure 7. The structure experiences a maximum spectral acceleration of 1.75 (1/g) at yield spectral displacement of 45mm (0.045m. The results of seismic demand model are presented in table 5.

Table 5. Analysis result of seismic demand model

Spectral Displacement S_d (m)	Spectral Acceleration S_a (1/g)
0	0
8.010×10^{-4}	1.518
1.602×10^{-3}	1.586
0.0320	1.700
0.0450	**1.750**

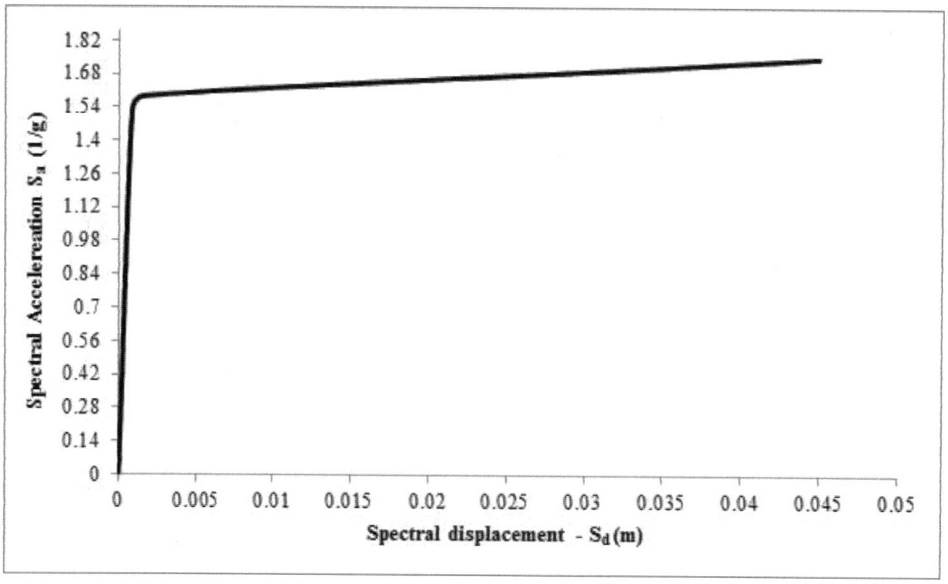

Figure 7. Bi-linear capacity curve

6.2. Damage state model

Comparing the building capacity curve, bilinear capacity curve and standard response spectrum data; the damage states defined for the structure chosen are as follows:

- Slight,
- Extreme,
- Moderate, and
- Collapse.

As per HAZUS [20], these damage states are characterized in structure chosen by the occurrence of damages. The deformed shapes of EFM corresponding to the damage states considered are represented in figure 8 to figure 10.

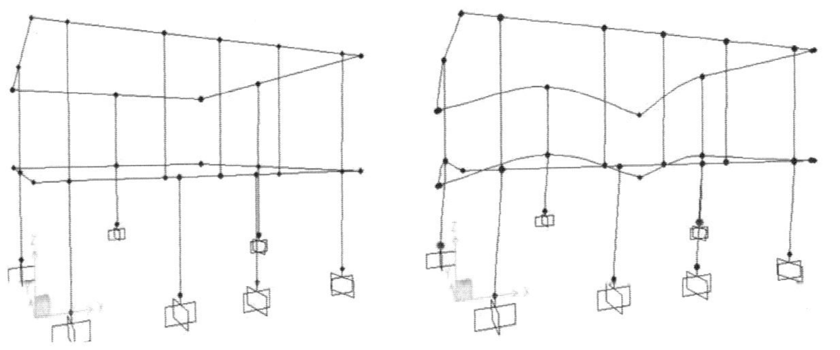

Figure 8. EFM and Deformed shape for slight damage state

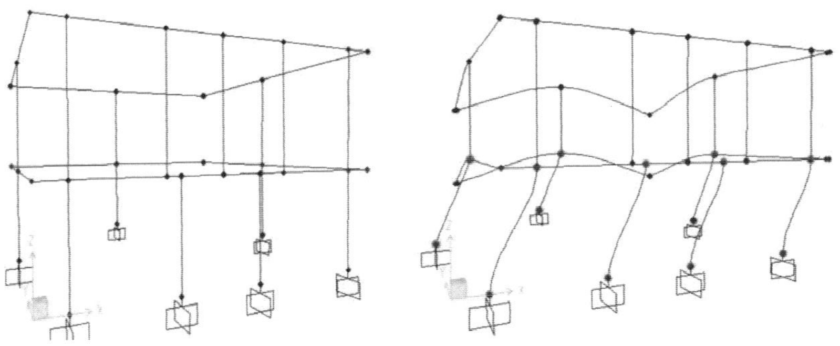

Figure 9. EFM and Deformed shape for moderate damage state

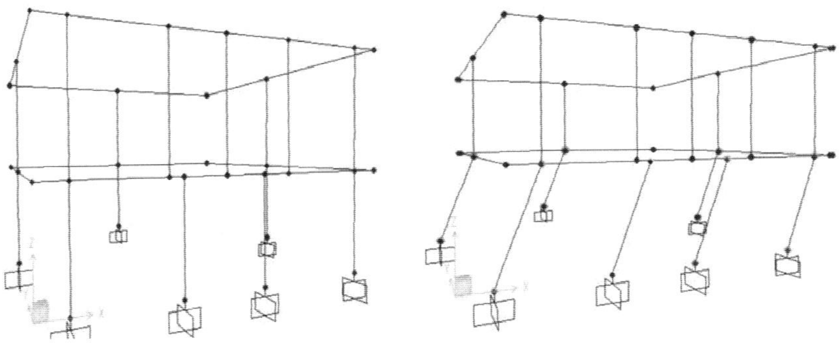

Figure 10. EFM and Deformed shape for extreme damage and collapse state

Based on the damages observed in the EFM analysed using NPOA method, the formation of damages are represented considering both building capacity and seismic demand of the structure. The damage state data of the structure studied is given in table 6, and the damage state model is represented in figure 11.

Table 6. Analysis result of damage state model

Base Shear (kN)	Roof Displacement (m)	Spectral Acceleration S_a (1/g)	Damage State Observed
0	0	0	-
2335.1542	8.010 x 10^{-4}	1.518	Slight
2434.5762	1.602 x 10^{-3}	1.586	Moderate
2645.3220	0.0320	1.700	Extreme
2650.3899	0.0450	1.750	Collapse

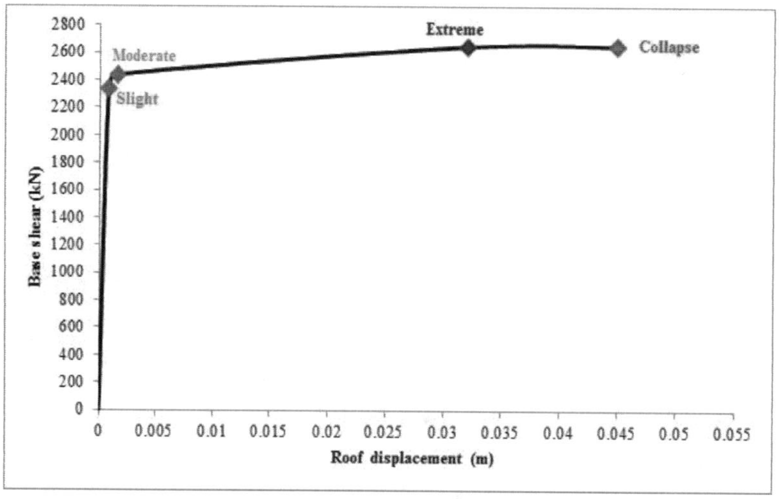

Figure 11. Damage state model

7. Fragility curve generation

As discussed earlier, fragility curve is defined as the probability of exceeding a given damage state defined as a function of earthquake loading [9, 13]. In this study, the function of earthquake loading considered is PGD. The Cumulative Distribution Function (CDF) defining probability of exceeding a specified damage state obtained as per HAZUS and Zentner, et. al., [14] is as follows:

$$P[ds|S_d] = \Phi\left[\frac{1}{\beta_{ds}}\ln\left(\frac{S_d}{S_{d.ds}}\right)\right] = \int_0^{S_d} \frac{1}{x\beta_{ds}\sqrt{(2\pi)}} e^{-\frac{1}{2}\left(\frac{\log(S_d/S_{d.ds})^2}{\beta_{ds}^2}\right)} dx \quad (2)$$

where, $S_{d,ds}$ is the median value of spectral displacement; β_{ds} is the standard deviation of natural logarithm of spectral displacement for damage state and Φ standard normal cumulative distribution function [15]. Since the demand spectrum for equivalent damping is dependent on building capacity, a convolution process is required to find the total variability β_{ds} as shown below:

$$\beta_{ds} = \sqrt{(CONV[\beta_c, \beta_D)^2 + (\beta_{\tau\, ds})^2} \quad (3)$$

where, $CONV\ [\beta_C, \beta_D]$ is a complex numerical process and difficult to perform as the demand and capacity are correlated.

HAZUS classifies URM structures as Un–Reinforced Masonry Medium rise buildings (URMM) and Un–Reinforced Masonry Low rise buildings (URML), depending on the height of the structure. The structural classification based on height of the structure as mentioned in HAZUS ® – MH 2.1 technical manual [16, 18] is given in table 7. Since the height of structure chosen for analysis is 7.925 m, it is considered as URMM.

Table 7. Classification of URM structures as per HAZUS

Building Notation	Height in feet	Height in meter
URML	≤ 15’	≤ 4.572
URMM	15’ to 35’	4.572 to 15.855

To avoid this difficulty, sets of pre-calculated valued of damage-state beta values have been compiled and provided in tabular form in HAZUS. The pre-calculated β_{ds} values for URMM structure of height 15.855m, as given in HAZUS are reproduced in table 8.

Table 8. Median and βds for URMM of height 15.855m as per HAZUS

Damage States	Median ($S_{d,ds}$) (in inches)	Median ($S_{d,ds}$) (in mm)	Beta (β_{ds})
Slight	0.50	12.7	0.99
Medium	1.01	25.7	0.97
Extreme	2.52	64.08	0.90
Collapse	5.88	149.352	0.88

The median and β_{ds} values for the structure chosen for analysis are calculated from the standard values given table 9 and considering the damage state model represented in figure 11, and these values for structure chosen is represented in table 9.

Table 9. Median and βds for URM structure analysed

Damage States	Median ($S_{d,ds}$) (in mm)	Beta (β_{ds})
Slight	0.801	0.062
Medium	1.602	0.060
Extreme	32	0.45
Collapse	44.9	0.265

Considering the building capacity, seismic demand and damage states; the probability of exceeding a particular damage state is obtained using the formula 2.

8. Results and discussion

Fragility analysis is performed on the structure modelled as per HAZUS ® - MH 2.1 for pre-code URMM structure, since the structure studied was constructed during early 20th century, exceeding a specified damage state is obtained using the formula discussed earlier. The fragility curves corresponding to four damage states i.e. slight, moderate, extreme and collapse damage states are given in figure 12 respectively. The combined fragility curve is represented in figure 13.

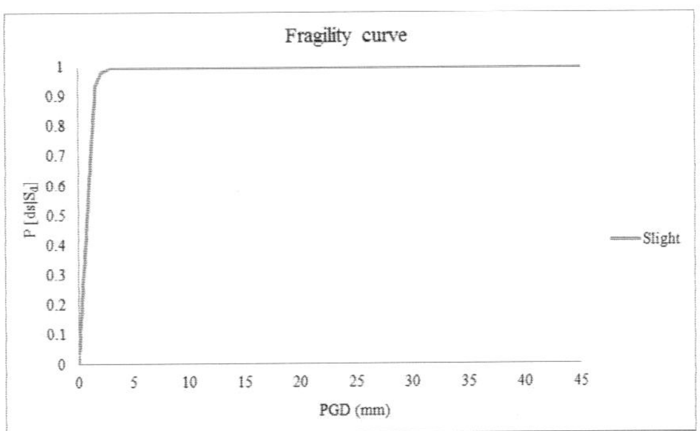

(a)

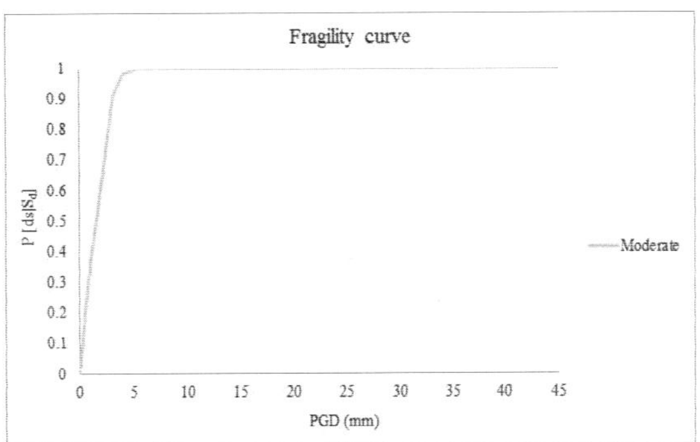

(b)

Figure 12. Fragility curves (a) slight & (b) moderate damage state

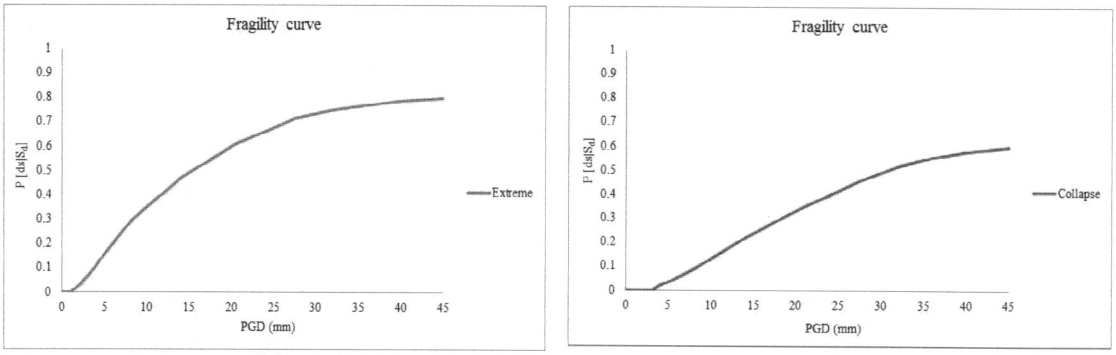

(c) (d)

Figure 12. Fragility curves (c) extreme & (d) collapse damage state

The fragility curves were derived using simplified log – normal distribution based procedure as per HAZUS considering four damage states. From figure 13, it is clear that slight and moderate damages occur at a very short interval after the occurrence of seismic action. The probability of occurrence of extreme damage is 80%, whereas the probability of structure to collapse is 60%.

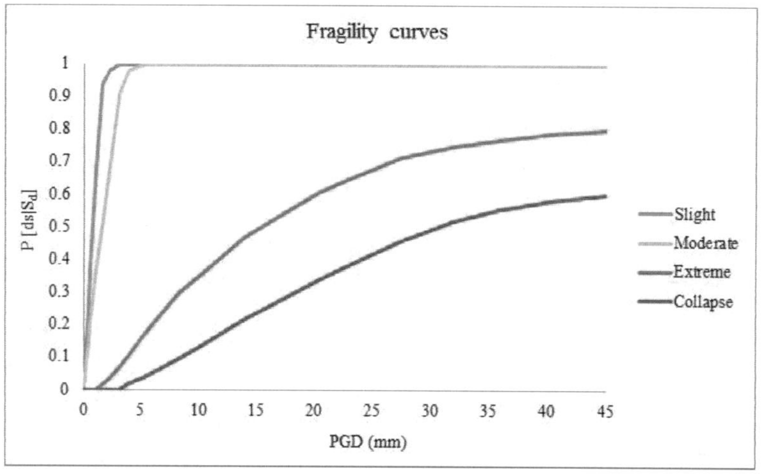

Figure 13. Fragility curves – combining all four damage states

9. Conclusion

A methodology for derivation of analytical fragility curves for URM buildings constructed during early 20th century, based on EFM and NPOA of the entire structure was discussed in detail. URM building investigated in this paper is a representative for structures constructed during early 20th century present in seismic zone – III (Chennai).

Probability of exceeding a specified damage state for slight, moderate, extreme and collapse damage states are 100%, 100%, 80% and 60% respectively. Formation of hairline cracks on walls, large cracks around door and window openings, movement of lintels, cracks at base of parapets were observed at earlier stage of seismic excitation. With increased seismic intensity, the structure fails/collapses due to out – of – plane failure of the wall [19]. The structure's damage probability is high at slight and moderate seismic intensity. The structure needs to be retrofitted for slight and moderate damages that are expected to occur.

References

[1] S Varadharajan V K Sehgal and Babita Saini 2013 Review of different Structural irregularities in buildings *J. Struc. Engg.* **39** 538-63.

[2] Anand S Arya 2008 Seismic Assessment of Masonry Buildings *J. South Asia Disaster Studies* **1** 23-43.

[3] K A Korkmaz 2009 Seismic safety assessment of unreinforced masonry low-rise buildings in Pakistan and its neighbourhood *Natural Hazards & Earth System Sciences* **9** 1021-31.

[4] Ricardo Leon Bonett Alex Barbat Lluis Pujades Sergio Lagomarsino and Andrea Penna 2006 Performance assessment for unreinforced masonry building in low seismic hazard areas *Enero – Junio de.* **5(8)** 105-18.

[5] R Sabatino and G Rizzano 2011 A Simplified Approach for the Seismic Analysis of Masonry Structures *The Open Construction & Building Tech. J.* **5** 97-104.

[6] Naida Ademovic Mustafa Hrasnica and Daniel V Oliveira 2013 Pushover analysis and failure pattern of a typical masonry residential building in Bosnia and Herzegovina *Engg. Struc.* **50** 13-29.

[7] Sergio Lagomarsino Andrea Penna Alessandro Galasco and Serena Cattari 2013 TREMURI program: An equivalent frame model for the nonlinear seismic analysis of masonry buildings *Engg. Struc.* **56** 1787-99.

[8] Guido Magenes 2000 A method for pushover analysis in seismic assessment of masonry buildings *Proc. 12th World Conf. on Earthquake Engg.* 1866.

[9] F. Colangelo 2008 On the computation of seismic fragility curves *Proc. 14th World Conf. on Earthquake Engg.* Beijing, China.

[10] Yogendra Singh Dominik H Lang J S R Prasad and Rajesh Deoliya 2013 An Analytical Study on the Seismic Vulnerability of Masonry Buildings in India *J. Earthquake Engg.* **17** 399-422.

[11] U Vipin Unnithan A Meher Prasad and B N Rao 2008 Development of fragility curves using high dimensional model representation *Proc. 14th World Conf. on Earthquake Engg.* Beijing, China.

[12] G M Calvi R Pinho G Magenes J J Bommer L F Restrepo-Vélez and H Crowley 2006 Development of Seismic Vulnerability Assessment Methodologies over the Past 30 Years *ISET J. of Earthquake Tech.* **43** 75-104.

[13] F Ranjbaran and M Hosseini 2012 Analytical Fragility Curves of Confined Masonry Buildings *Proc. 15th World Conf. on Earthquake Engg*, Lisboa.

[14] Amin Karbassi and Pierino Lestuzzi 2012 Fragility Analysis of Existing Unreinforced Masonry Buildings through a Numerical-based Methodology *The Open Civil Engg. J.* **6** 121-30.

[15] Thomas M Frankie, S M ASCE Bora Gencturk A M ASCE Amr S Elnashai and F ASCE 2013 Simulation-Based Fragility Relationships for Unreinforced Masonry Buildings *J. Struc. Engg,* **139** 400-10.

[16] Elisa Lumantarna Jerry Vaculik Mike Griffith Nelson Lam and John Wilson 2006 Seismic fragility curves for un-reinforced masonry walls *Proc. Conf. Earthquake Engg.,* Australia, Canberra.

[17] J Ruiz-Garcia A Teran-Gilmore and Oscar Zuniga-Cuevas 2010 Simplified drift-based fragility assessment of confined masonry buildings *Proc. 9th U.S. National and 10th Canadian Conf. on Earthquake Engg.,* Toronto, Ontario, Canada.

[18] Dimova S L Hirata K 2000 Simplified Method of Fragility Analysis of Structures with Non-Traditional Seismic Protection *Proc. 12th World Conference on Earthquake Engg.,* Auckland, New Zealand.

[19] Edward L Wilson 2001 *Three – Dimensional Static and Dynamic Analysis of Structures, A Physical Approach with Emphasis on Earthquake Engineering*, Computer and Structures Inc, Berkeley California USA.

[20] Department of Homeland Security *HAZUS ® - MH 2.1 Technical Manual* Federal Emergency Management Agency Washington D.C.

Polluted Landfill adaptation into agricultural soil: heavy metal phytoremediation with Indian Black Mustard (Brassica Nigra) and Dolomitic Lime Fertilizer.

C HariharaSudhan[1], B Anuradha[2], S Jayadevsivani[3] and K Gokul[4]
[1]Professor, Department of Civil Engineering, Chennai Institute Technology, Chennai - 600069
[2]Assistant Professor, Department of Civil Engineering, Chennai Institute Technology, Chennai -600069
[3]UG Student, Department of Civil Engineering, Chennai Institute Technology, Chennai – 600069
[4]UG Student, Department of Civil Engineering, Chennai Institute Technology, Chennai – 600069

hariharasudhanc@citchennai.net[1],anu_bas2003@yahoo.com[2],jayadevsivakumar55@gmail.com[3],gokulsknithi7@gmail.com[4]

Abstract. Bio-medical and electronic waste accumulation from industry and anthropogenic sources is deposited at landfill sites over the years, possessing waste degradation into heavy metal deposition in soil such as Cd, Ni , Pb, which is detrimental to the developing embryo development and affects respiratory function. The hyper-accumulator herb, such as Indian mustard (Brassica Nigra), which was able to absorb heavy metals from the soil and water and wipe out soil after a period of time and disposed of it. Dolomitic Lime Fertilizer is a Inorganic compound derived from mixture of calcium carbonate ($caco_3$) and magnesium carbonate ($mgco_3$) are used for the sustained agriculture. The Extraction of heavy metals along with Dolomitic lime helps to increase the nitrogen fixation rate in soil and reduce the acidity nature of the soil also acts as a soil conditioners. They provide eco-friendly, cheap and renewable resource for the subsequent years. Phytoremediation of contaminated sites is requires low cost and more renewable compared to remediation strategies that involve removing polluted soil constituents.

Keywords: Phytoremediation, Heavy metals, Dolomitic lime, Indian mustard.

1. INTRODUCTION

Soil heavy metal pollution is an emerging environmental controversy that affects both agricultural productivity and irrigation systems throughout the world. Such heavy metals and metalloids possess an immense amount of bio-hazardous material such as Cd, Ni , Pb and Cr are known to be non-essential heavy metals that are hazardous due to concentration beyond their acceptable limit. The higher concentration of non-essential minerals causes living organisms to malfunction. Via natural calamities and human anthropogenic activities these metals join the Soil environment in their everyday lives. In order to innovate the research and experimental study on Indian Black Mustard (Brassica nigra), we carried out a detailed study to enrich the contaminated soil using inorganic wastes and Starch water as a bio fertilizer to reduce the level of PH in the soil by producing organic acids, thus reducing the existing microbial organisms in the soil and protecting the soil texture. The presence of heavy metal compounds from the soil enhances the leachate toxicity that penetrates the soil surface and dissolves in the residential areas surrounding the dumping yard at the underground water table. Thus it offers an environmentally friendly conversion of heavy metal contaminated landfill for plant purposes to enrich agricultural soil.

2. REVIEW OF LITERATURE

Chhotu D. Jadia and M.H.Fulekar (2008), the main aim of the study is to promote the Phytoremediation process to clean up the heavy metal contamination from the environment. It is also related to the In situ, solar driven technology to make use of vascular plants to absorb and transport the metals from soil to shoot system. They applied their technology in shallow soil, ground water and surface water bodies to provide a eco-friendly remedial measures.

M.Ghosh and S.P Singh (2005), this project is about study of phytoremediation and its applications. They study the no-edible species of plants which not included in food chain to remediate soil .They propose the in-situ and ex-situ method of cultivations of hyper accumulator plants to uptake heavy metals from soil. They study natural ability of the plants and its function on this process.

H.G.Aliyu and H.M.Adamu (2014), this study shows that, Maize can cleanup contamination in very short period of time without any cost implications and no time consumption. They suggested that temperature and light not only influcing the growth, but also affects the metal absorption. The temperature, water control, competing ions are parameters affects the absorption rate of heavy metals by the plant species. Hazrat Ali, Ezzat Khan and Muhammad Anwar Sajad (2013), They proposed various techniques involved in phytoremediation such as phytoextraction, phytovolatisation, Rhizofiltration, Phytomining etc .They also study the mechanism and role of phytochelatins and metallothioneins in phytoextraction.

Amin Mojiri (2011),A study was carried out to investigate the ability of Corn (Zea mays) plant for phytoremediation. At the end of 60 days, they analysis the soil texture, cation exchange, Ph, electrical conductivity and extractable cadmium and lead are measured after the test. It shows that Corn is a effective accumulator plant for phytoremediation of cadmium and lead from the contaminated soil.

Scott Angle and Alan JM Baker (1997), they conclude that hypertolerence is the basic to hyper accumulation and high rates of uptake in plants. They increase the annual rate of phytoextraction by improved hyper accumulator plants and agronomic technology. It uses "Metallophores" to aid phytoextration of soil metals.

Wendy Ann Peer and Angus S.Murphy (2015), they employed technique such as EDTA chelators to boost up the rate of uptake of metals in Indian mustard and sunflower. They analysis all non-essential heavy metals such as Cd, Ni , Pb ,Cr ,Hg, Br etc.. Some plants selected to uptake high concentration of metals such as gold and Ni.

I.D.Pulford and C. Waston (2002), they introduced the concept of phytoremediation of metals using specific trees. It controls the movement of heavy metals at Rhizosphere of the trees. It accumulates the high amount of heavy metals for more number of years. Field trials shows that cleanup of such soils could be achieved within a few years.

3. METHODOLOGY

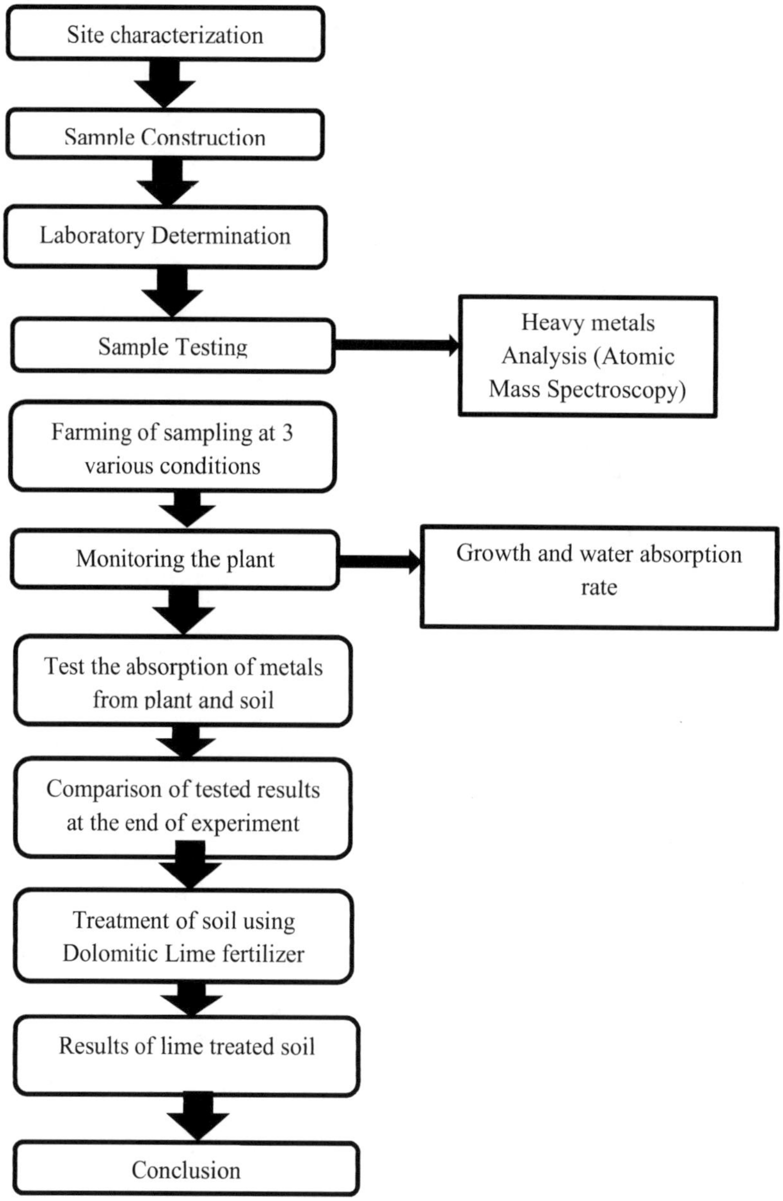

4. MATERIAL AND METHODS

4.1. Site characterization

The population of the city of Chennai is 43.43 lakhs, according to census 2016, and the average per capita of solid waste generated from City was estimated at 585 grams. Chennai's cooperation claims that every day, it produces 5000MT of garbage. They are collected the waste through different methods of collection and transported to Kodungaiyur dump yard. This started in 1987. Currently over 26 acres of plot in this area is used for garbage dumpsite, while only 100 acres of land was used to dump hazardous waste. Kodungaiyur's latitude and longitude will be 13 ° 08'02 "N, and 80 ° 16'09" E. Elevation of the site lies above the mean sea level of 7 metres.

4.2. Sample Construction

The experiment is performed under room temperature and optimum pressure at open sunlight. Soil samplings from site depth were taken from 0-30 cm. Soil samples in experimental atmosphere are allowed to dry for 24 hours at a temperature range of 25-30o C. To extract hard materials from the soil, the samples are sewn by 12mm sieve, then the soil samples are subjected to 4.25 mm sieve. The mustard plantlets are planted in a glass container filled with a mixture of sieved samples and e-waste (including other inorganic compounds) collected in Two Layers from the site. The experiment consists of three samples each weighing 4.81 kg, the soil sample density is approximately 2082kg / m3, and the heavy metal contamination in the samples is cadmium (0.10 mg / kg), nickel (42.40 mg / kg), and lead (97.80 mg / kg). Upon seed planting the water is irrigated as per requirement. Samples were taken, after 50 days, for processing.

4.3. Laboratory determination

Until titration, the conical flask is washed clean with acid and distilled water. 1 g of each of the soil samples is put in this tank. The concentrated acids such as 25 cm_3 HNO_3, 6 cm_3 H_2so_4, 5 cm_3 per chloric acid are mixed together and subjected to furnace. Where the solution is gently mixed at medium temperature, and where standby. Thus the procedure continues until the expulsions of the white dense flumes.45cm3 of deionized water is added for a minute to the solution within the furnace condition. It is then cooled at low temperature and filter for 5 minutes. The filtered solution is subjected to heavy metal analysis using Atomic mass spectroscopy.

5. SAMPLE TESTING

From atomic absorption spectrometry results,

The soil is polluted with a high toxicity content relative to the allowable limits of metals concentration in soil as given by TCLP 's provisional code-Toxicity Characteristic Leaching Procedure is a test to assess the mobility of pollutants in solid waste or soil. And WHO (1996); Housing Ministry, the Netherlands, 1994.

Table 1. Concentration of Heavy Metals from the Sample from the Site

S.NO	PARAMETERS	METHODS	UNITS	RESULTS
1.	Lead	TNTH/SOIL/SOP/022	mg/kg	99.6
2.	Cadmium	TNTH/SOIL/SOP/022	mg/kg	AQL(LOQ:0.8)
3.	Nickel	TNTH/SOIL/SOP/022	mg/kg	43.23

AQL-Above Quantification Limit and LOQ- Limit of Quantification

6. MONITORING OF SAMPLES

Table 2. Increase in Length of the Plant in Centimeters (cm)

Day	Length in cms			Average length in cms
	sample A	sample B	sample C	
0	0	0	0	0
8	3.36	3.25	3.06	3.12
16	9.6	9.2	8.7	9.26
24	17.56	17.2	16.9	18.22
32	32.87	30.5	31.7	29.6
40	56.87	52.8	55.4	50.02
44	60.44	60.12	59.67	58.07

The growth of the samples after 40 days.

7. TEST THE ABSORPTION OF METALS FROM PLANT AND SOIL

The plants from the samples are plucked out together with the roots being washed and dried for 48 hours under sunlight. It is then converted to fine-powered form. It undergoes a Spectrometry of Atomic Absorption.

Table 3 Concentration of Heavy Metals after End of Experiment

S.NO	PARAMETERS	METHODS	UNITS	RESULTS
1.	Lead	TNTH/SOIL/SOP/022	mg/kg	17.80
2.	Cadmium	TNTH/SOIL/SOP/022	mg/kg	AQL(LOQ: 0.75)
3.	Nickel	TNTH/SOIL/SOP/022	mg/kg	10.43

BQL-Below Quantification Limit and LOQ- Limit of Quantification

8. TREATMENT OF SOIL USING DOLOMITIC LIME

8.1. Dolomitic Lime

Dolomitic lime is the additional nutrient compound which acts as the Agricultural lime; it supplies magnesium and Calcium at the ratio of 6:1.It suppress the acidity of the treated soil and maintain the PH balance over the period. It also act as the disinfectant in farming to resist the growth of bacteria and other microorganisms without causing any damages to the livestock. Dolomitic Lime Fertilizer is an Inorganic compound derived from mixture of calcium carbonate ($caco_3$) and magnesium carbonates ($Mgco_3$) are used for the sustained agriculture. The Extraction of heavy metals along with Dolomitic lime helps to increase the nitrogen fixation rate in soil and reduce the acidity nature of the soil also acts as a soil conditioners. The optimum range of PH in agricultural soil varies from 5.5 to 6.5.If the soil exist high range of acid content such as 4.5 to 5.0, it may cause deficiency of nutrient to the crops and damage the vegetation. The Dolomitic lime acts as the conditioner and reduces the soil acidity up to 80% and helps in Nitrogen Fixation in plants.

8.2. Experimental Procedure

1. For applying lime to the sample requires 60ml of lime powder for 1 square feet area of the soil.

2. Sprinkle the lime over the dry area of the sample at the moderate temperature.

3. Dig the sample up to 10cm and pour the lime, mixed with water of ratio 1:3.

4. After a week, tilt and spread the soil samples and exposed to open atmosphere before the sampling of seeds.

5. Then the soil may undergo for Vermi-Composting.

9. RESULTS AND ANALYSIS

9.1 Comparison of end results with the samples.

Table 4. Results of Experiments

S.NO	HEAVY METALS	CONTROL	CONTAMINATED SOIL
1.	Pb	17.80	96.80
2.	Cd	0.75	0.8
3.	Ni	12.43	43.23

The concentration units of above metals are in mg/kg.

From the above findings, after cultivation of mustard plantlets at the end of 40 days, there is a progressive decrease in the concentration of heavy metal content in soil. Thus, it demonstrates that Indian Black mustard has a remarkable capacity to absorb and store high soil metal concentrations. The ability of different plant species to absorb and store heavy metals in their tissue is commonly known as hyper accumulators. It has the ability to withstand the conditions and hardness of hard soil. The harvested soil after 40 days shows that only in ideal temperature conditions and sufficient amount of water supplied to the samplings the plants expand and pick up the heavy metals. Many non-essential metals like mercury, arsenic, chromium, etc. ... are absorbed Compared to Ni, Cd and Pb, at a minor level.

9.2. Results of Dolomitic lime Fertilizer

- High yield obtained by correcting the PH and neutralizing the soil condition.
- Increases the Water soluble rate of soil and provide efficient phosphate content to soil.
- It resists the degradation of Molybdenum amount in field.
- Enhances the growth of earthworms and fix the organic material in soil such as nitrogen, sulphur and phosphorus.
- Preventing the accumulation of heavy metals.
- Maintain the Structure of soil and its Texture.

10. CONCLUSION

The Indian black mustard plant has the ability to remove heavy metal constituents from landfill soil using Phyto-extraction method as a result of the experiment. The findings show that at the end of the experiment there is a drop in concentration of Cadmium, Nickel and lead from the samples. The treatment of Starch water prepared from the Rice in fermented condition acquired with Lactobacillus microbe is used to purify the unwanted microbes from organic matter and also increases the plant's root's Nitrogen Fixing Capability (NFC). It also removes the foul odor of organic matter present inside the soil during decomposition. It retains the soil's original Habitat and increases crop yield.

11. REFERENCE

[1] Amin Mojiri 2011 *The potential of Corn (zea mays) for Phytoremediation of Soil Contaminated with Cadmium and Lead, Journal of Biological and Environmental sciences* **Vol.5(13)** pp.17-22. https://www.researchgate.net/publication/225285964

[2] Chhotu D.Jadia, and M.H.Fulekar 2009 *Phytoremediation of heavy metals: Recent techniques African journal of Bio-technology* **Vol.8 (6)** pp.921-928. http://www.academicjournals.org/AJB

[3] Denneman and Robberse, 1990 *WHO permissible limits for heavy metals in plant and soil Ministry of Housing Netherlands* https://www.omicsonline.org/articles-images/2161-0525-5-334- t011.html

[4] Fang-Chih Chang, *et al 2014 Phytoremediation of heavy metal contaminated soil by Jatropha curcas. Ecotoxicology C.-H. Ko Bio energy Research Center* **Vol.23** pp1969-1978

[5] H.G.Aliyu and H.M.Adamu 2014 *The Potential of Maize as Phytoremediation tool of heavy metals. European scientific journal* **Vol.10 (6)** pp1857-7431.

[6] Hazrat Ali,Ezzat Khan, and Muhammad Anwar Sajad *2013 Phytoremediation of heavy metals- Concepts and applications Chemosphere-Environmental Chemistry* **Vol.91(7)** pp869-881. http://dx.doi.org/10.1016/j.chemosphere.2013.01.075

[7] M.Ghosh and S.P.Singh 2005 *A Review of Phytoremediation of Heavy metals and Utilization of It's by Products.Asian journal of Energy and Environment* **Vol.6 (04)** pp214-231. http://www.asian- energy-journal.info

[8] Mohammad Iqbal LONE *et al* 2008 *Phytoremediation of heavy metal polluted soils and water: Progresses and perspectives Journal of Zhejiang University* **Vol.9(3)** pp210-220. http://**doi:**10.1631/jzus.B0710633

[9] P. Ahmadpour et al 2012 *Phytoremediation of heavy metals: A green technology. African Journal of Bio technology* **Vol. 11(76)** pp14036-14043. http://www.academicjournals.org/AJB

[10] Revathi.K, Haribabu.T.E, and Sudha. P.N 2011 *Phytoremediation of Chromium contaminated soil using Sorghum plant International Journal Of Environmental Sciences* **Vol.2(2)** pp429-440. https://www.researchgate.net/publication/284690921

[11] Rufus L Chaney *et al 1997 "Phytoremediation of soil metals" Current Opinion in Biotechnology* **Vol.8 pp279**-284. http://bomednet.com/elecref/0958166900800279

[12] *Toxicity characteristic leaching procedure.U.S Environmental Protection Agency (EPA)* 1992 Document no **.Sw-846.** https://www.omicsonline.org/open-access/heavy-metal-concentrations-in-plants-and-soil-along-heavy-traffic-roadsin-north-central-nigeria-2161-0525-1000334.php?aid=63825

[13] Vimal Chandra Pandey 2012 *Phytoremediation of heavy metals from fly ash pond by Azolla caroliniana", Ecotoxicology and Environmental Safety* **Vol.82** pp.8-12 http://dx.doi.org/10.1016/j.ecoenv.2012.05.002

[14] Wendy Ann Peer *et al* 2002 *Phytoremediation and hyper accumulator plants" Research gate* http://www.hort.purdue.edu/hort/research/murphy/pdfs/metals11.pdf

IOP Conf. Series: Materials Science and Engineering 1026 (2021) 012006 doi:10.1088/1757-899X/1026/1/012006

Study On Leachate And Its Effects On Soil Properties

Damaraju Lakshmi Lavanya[1]
lldamaraju@gmail.com
lavanya.damaraju@vbithyd.ac.in
Assistant Professor , Department of civil engineering
Vignana Bharathi Institute Of Technology

Abstract:
The resources of earth are supporting the man kind , but various actions of man increases the generation of waste causing a huge solid waste , presently which is a third pollution in world –wide . The gradual decomposition of solid waste , includes various bye-products and soluble materials like nutrients , pesticides ,chemicals are leaching into the soil and causing a soil pollution as well as ground water pollution .The leaching liquid known as Leachate, percolates into the layers of soil and effects the Various Geo-technical properties. The study confines to identify the effects of the leachate on soil properties and also to study the variations in treating the leachate by Recirculation method to decrease the concentrations levels . By collecting the waste, an artificial dump yard is created and water is added to the waste for formation of leachate. A setup is made, the leachate is treated by one of the Leachate treatment methods the method of recirculation. Leachate is collected and parameters pH, BOD, COD, Alkalinity ,Sulphates, Nitrates, Conductivity, Total hardness, Total Dissolved Solids, Total Suspended Solids, Chlorides of leachate are tested in before and after the recirculation process. The soil property at the site was observed before and after leachate formation. Also the recirculated leachate is sprinkled in the site, soil properties was tested. For soil engineering properties, geo-synthetics like geo-membranes can be used at the dumping site to reduce the effect of leachate on soil properties.

Keywords : Solid Waste, Leachate , Recirculation ,Soil Properties ,Geo-membranes

1. Introduction :

The generation of solid waste shows a considerable variations in urban and small towns . An average of 450 grams of waste will be generated per person per day . In urban places , with all modern facilities an average percapita of 650 grams will be generated . Some types of waste are consisting of materials which can decompose with the available moisture and rainfall like food waste, paper , chemicals etc . .Utilization of open lands or Dumping of Municipal solid waste is a major concern. Expanding city lines increases for the outlook of land availability Rapid development of the city outskirts and scarcity of land in the city creating the issue of dump yard stations and also reclaiming the dumpsite for future development like constructing the buildings. Open dump yards and Landfills are the conventional methods for solid waste disposal where large area of open –spaces are available. Landfilling method is causing environmental problems such as odorous smell, infectious vectors, leachate pollution etc .Leachate is a liquid, arises from the municipal solid waste

decomposition process and percolates into ground water and causes Ground water pollution. Leachate percolation ,the impurities present in the Leachate Liquid causing a potential threat by affecting the quality of ground water and also causes soil pollution , some effects in the properties of soil . This changes effects the constructing activities, like load carrying capacity of the soil, causing increasing the cost of Construction with by applying different remedies .
The present study observes the variations in leachate characterization and treated the Leachate by recycling or recirculation method

2. Methodology :

The study was conducted by Collecting the waste from municipal waste dump and prepared a artificial dump yard, to study the Leachate and soil properties before and after the generation of Leachate .
The available soil at the site is murram soil , two artificial dumping yards was created .Each pit is of 1.5 X1.5 m. The pit is dug at a depth of 15 cm. One pit is left simply and on the other geomembrane is laid. In each pit around 50 kg of waste is placed .Based on rainfall data in the study location, water is poured. Daily 3 liters of water is poured into the two dumping yards. The 3 litres of water was decided based on the rainfall data at the selected site. This process is continued for 14 days. Before, the soil properties of the fresh soil are tested and values are noted. After 14 days, the leachate will be formed and soil properties are tested for both the pits by removing the waste from it. The values obtained from tests are noted.Also the soil properties was tested collecting the soil sample where the Leachate percolation is done in the soil sample

Figure 1: Artificial dump yard set up

2.1 Artificial leachate formation

- Artificial leachate is prepared by an experimental setup. This setup consists of
  - Pipe - length(2m),diameter(16cm)
  - Dummy
  - Stand
  - Collecting tubs
  - Coarse aggregates

  A layer of coarse aggregate was laid at the top and bottom of solid waste to retain suspended solids to pass through. The pipe is held in vertical position with the help of stand. Water is added to waste of 1 litre per day for 14 days to generate the leachate . The leachate produced after adding of water for 14 days is 9 litres. 2 litres of leachate is taken separately for testing the parameters of leachate. Remaining 7 litres is taken for recirculation. In re-circulation

IOP Conf. Series: Materials Science and Engineering 1026 (2021) 012006 doi:10.1088/1757-899X/1026/1/012006

process fresh leachate is added into the pipe columnof 1 litre per day for a week .Here, the fresh leachate reacts with the waste present in the pipe The leachate produced after recirculation process is of 4.5 litres. 1 litre is taken for tests and remaining is poured on to the fresh soil at the study area. Then the soil properties are tested after pouring of recirculated leachate on to the study area. All values are noted and calculated.

Figure 2: Artificial Leachate Formation

Figure 3: Recirculated Leachate

Figure 4: Fresh Leachate

2.2 GEOMEMBRANE

A geomembrane is very low permeability synthetic membrane liner used to arrest the movement of the fluid in a material . Geomembrane are made from relatively thin continuous polymeric sheet. It acts as a barrier between the any two layers .The Lifetime of the Geo-Membranes in landfill can be approximately used for about 25 years . The different Geo-membranes is having different life times in usage

Figure 5: Geo-membrane used in the study

Specifications:

- Gsm-150 (gram per square metre)
- Smooth surface
- Thickness – 2.5mm
- Impermeable
- Longevous(30 years)

Characteristics:

- Low biological degradation
- High resistance to chemicals
- High impact resistance
- High tear resistance
- High Interface Shear strength

2.3 Tests Performed on soil and Leachate

The following are the tests Performed on the soil

- California Bearing Ratio test
- Direct shear test
- Permeability test

California Bearing Ratio Test:

To evaluate the strength of below sub - layers of the soil, this test is conducted .The CBR is a measure of resistance of a material with respect to the load.

Direct Shear Test:

To identify the shear strength of the soil properties and is function of Normal load, angle of friction which determines the interlocking between the soil particles

Permeability Test:

Constant Head Method:

The test was done to determine the permeability Nature of the soil , to calculate the Discharge in a given time .

Presently this test was conducted to Known the rate of movement of Leachate into the soil .
The Various tests conducted on to study the Characteristics of Leachate

- pH
- BOD
- COD
- Alkalinity
- Sulphates
- Nitrates

IOP Conf. Series: Materials Science and Engineering 1026 (2021) 012006 doi:10.1088/1757-899X/1026/1/012006

- Conductivity
- Total hardness
- Total Dissolved Solids
- Total Suspended Solids
- Chlorides

3. Results And Observations

The soil samples was collected and tested before the formation of dump yard, after the formation of leachate and with recirculated leachate. The different tests were conducted on freshly formed leachate and recirculated leachate. The rate of generation of Leachate in recirculation was observed as decreased by 15 ml

Table 1: soil sample Tests results

Tests	Fresh soil	Soil sample after formation of leachate without geomembrane	Soil sample after formation of leachate with geomembrane	Soil sample with recirculated leachate
Direct shear test	cohesion=30 Angle of internal friction=10.42^0	cohesion=75 Angle of internal friction=2.1^0	cohesion=50 Angle of internal friction=8.75^0	cohesion=45 Angle of internal friction=3.9^0
Permeability	0.089mm/sec	0.08mm/sec	0.085mm/sec	0.083mm/sec
CBR test	20.56%	10.6%	16.78%	12.09%

Table 2: The following are the Tests conducted on Leachate

Tests Performed	Fresh leachate	Recirculated leachate
pH	6.5	7.48
Conductivity	6.59	7.49
TDS	10208 mg/l	8964 mg/l
TSS	56 mg/l	45 mg/l
Chlorides	282 mg/l	220 mg/l
Alkalinity	3926 mg/l	2897 mg/l
BOD	1023 mg/l	912 mg/l
COD	3223 mg/l	2987 mg/l
Sulphates	120 mg/l	90 mg/l
Nitrates	35.2 mg/l	29.6 mg/l

4. Conclusions:

1) The recirculated leachate parameters such as Alkalinity, Nitrates, Chlorides etc, are decreased as compared with fresh leachate.
2) By leachate recirculation process the effects of landfill leachate can be decreased.

3) As compared with fresh soil properties and dumping yard, the parameters like shearing strength, permeability is decreased by different percentages with leachate percolation process.

4) Usage of geo-membrane at the dumping yard site does not show any decrease in the soil characteristics and it will be as same as with fresh land.

5. References

1) ASTM D4439 Standard Terminology for Geosynthetics. ASTM International.

2) Lee, A. H., Hamid, N. and Hung Y. T. **(2010)**, Influence of Waste Age on Landfill Leachate Quality. *International Journal of Environmental Science and Development,* **1(4).**

3) Soong, T.-Y., Lord, A.E., and Koerner, R.M**., (1994**), "Stress Relaxation Behavior of HDPE Geomembranes", *Proceedings of the Fifth International Conference on Geotextiles, Geomembranes and Related Products, Southeast Asia Chapter of the International Geotextile Society,* Singapore, pp **1121-1124.**

4) Soong, T.-Y, and Koerner, R.M., **(1997),** "*Behaviour of Waves in High Density Polyethylene Geomembranes: A Laboratory Study", Proceedings of the* ***11th*** *GRI Conference, GRI, Folsom, PA, USA,* **pp 128-151.**

Websites:

1) Atlas of Groundwater in Virginia: http://capp. water.usgs.gov/gwa/ch_l/index.html

2) https://en.wikipedia.org/wiki/Geomembrane

3) http://www.springer.com/engineering/civil+engineering/journal/**40891**

4) http://ascelibrary.org/journal/jggefk

IOP Conf. Series: Materials Science and Engineering 1026 (2021) 012013 doi:10.1088/1757-899X/1026/1/012013

Soil Stabilization Techniques for Roads on Expansive Soils and Loose Sand

L Kanagarathinam[1], D Anburaja[2], G Sudharson[2], P Keerthana[3] and S Karpoorajothi[3]

[1]Associate Professor, Department of Civil Engineering, Saveetha Engineering College, Chennai-602105

[2]Assistant Professor, Department of Civil Engineering, Saveetha Engineering College, Chennai-602105

[3]UG Student, Department of Civil Engineering, Saveetha Engineering College, Chennai-602105

Corresponding author: kanagarathinaml@saveetha.ac.in

Abstract. There are various techniques to stabilize soil which changes in volume due to vibration. Structural foundation in regions with expansive soil or loosely filled soil have to be laid after stabilizing such soils or by adopting the depth of foundation below the zone of movement. The tunneling induced vibration or alternative swelling and shrinking of soil due to seasonal change or change in ground water table level, causes movement of soil. This movement of soil result in additional stresses on the foundation, if the foundation depth is within this zone of movement. Normally, this zone of movement will be 1.5 to 2m below ground level. In some cases, this zone of movement may extend up to 5 to 6m. In this paper we have reviewed the various methods of stabilizing the expansive soil and loosely filled soil, especially the soil subgrade below the roads. Moreover, tunneling induced vibration due to underground metro rail construction also affects the soil behavior which in turn affects roads and nearby structures. Hence we have studied the effect of vibration on sandy soil by conducting laboratory experiments in the vibration table . The experimental result showed that the sandy soil settled more due to vibration when compared to the soil stabilized with gravel column. Addition of geomaterials such as gravel/stone in the form of column reduced the settlement level of loosely packed sand. Hence installing stone columns in soil is a better method to reduce settlement due to vibration that are produced during tunneling operations. Stone columns also prevents pore water pressure that builds up during earthquake vibration.

1. Introduction

Tunneling for road and rail networks is inevitable as it saves space in urban cities in developing countries and cities are developed rapidly for the purpose of modernization. Increasing population and technology advancement demands improvement of road and rail network in major cities. Construction of road and railway lines requires vast area of land including its future expansion work. Due to non-availability of lands, we need to construct tunnels. There are two major problems encountered during the process of tunneling, they are vibration and settlement. Tunneling creates vibration, in turn nuisance to surrounding population, disturbance of working condition of sensitive devices, destroy the serviceability and durability of nearby buildings and structures.

The frequency level of ground and structural vibration depends on the source, soil-structure interaction, distance from the source and the condition of existing structures surrounding the construction site. Tunneling causes vibration and settlement of ground, hence it affects the stability of nearby buildings and structures. Mitigating the effect of vibration and settlement on nearby buildings, is the need of the hour. Granular wall or gravel drain has been effective in reducing the liquefaction effect and settlement.

Experimental comparative study have been carried out to understand the vibrational effect and resulting settlement. The settlement of soil before and after the implementation of mitigation technique is compared in this project. The main objectives of the project to measure the settlement of the natural river sand before and after stabilizing with geomaterial in the vibration testing machine. Then to compare both the settlement and to adopt suitable technique to reduce vibration induced settlement in sandy soil.

2. Literature review

Boominathan et al. (2019) studied the installation of gravel drains around the tunnel accelerates the pore pressure dissipation from the vicinity of the structure and there was a considerable reduction in the pore pressure accumulated at the invert of the structure .So we are using this process in our experimental study. Xie et al. (2018) discussed that the process of blasting operation, the shock waves, and earth quack waves subsequently developed. Due to the existence of these waves, the internal structures of the rock mass are inevitably influenced. Deckner (2017) developed a simpler and accurate method for estimating vibration dissipation as the distance from pile driving increase and predicting where ground settlement will occur.

Huang and Shi (2015)-The main design restriction for tunnel excavation in urban area is the induced ground deformation when the tunnel line passes near an existing building. In this paper, a numerical study was carried out to evaluate the possibility of reducing ground movements involved by tunnelling by pre-installing a passive protective barrier between the tunnel and the building exposed to potential damages. Liu et al. (2018) performed parametrical analyses using the Plaxis 2D and 3D Finite Element to define the soil-structure interaction mechanism that allow the reduction of ground movements induced by tunnel excavation. Diaphragm walls pre-installed between the tunnel and the buildings show to be effective in mitigation the effects induced by tunnelling.

3. Materials used for this study

The construction of geomaterial column in loose sand deposit is simulated in laboratory by installing cylindrical gravel column in the middle of mould filled with river sand.

3.1. Natural River Sand

Natural river sand is the most preferred choice as a fine aggregate material. River sand is the product of natural weathering of rocks over a period of millions of years It is mined from the river beds. The natural river sand are deposited in loose condition with more air voids. Construction of buildings near the river or in river basin pose serious threat as the loose soil tends to densify and cause settlement under mechanical vibrations or liquefy during earthquake vibrations. Pile driving in river sand vibrates the surrounding soil and stresses the nearby building. In this study, natural river sand is used to simulate loose soil condition.

Properties of Natural river sand

- Coarse sand (4.75-2.00mm) - 6.6 % by weight
- Medium coarse sand (2.00-0.475mm) - 73.5% by weight
- Fine sand (0.475—0.075mm) - 19.8% by weight
- Specific gravity - 2.67

- Bulk density (kN/m^3) - 14.5
- pH - 8.66

3.2. Geomaterial

Geomaterial can be used as a part of mix in concrete. They are also used in the slab and vapour barrier. They are used in the construction of driveways and roadways. In this study, aggregate of size 4.75mm passing is considered as geomaterial.

Properties Of Geomaterial

- Size of the aggregate - 4.75mm
- Fine modulus (%) - 2.864
- Grading zone - 2
- Bulk density(kN/m^3) - 17.42
- Percentage of voids (%) - 56.169
- Maximum percentage of bulking (%) - 18.3
- Corresponding moisture content (%) - 2
- Water absorption (%) - 2.14.

4. Experimental Procedure

The experiment was carried out in the Cushioned Steel Vibration table 75 * 75cm, 3600 r.p.m., 440V. Empty weight of the cylindrical mould was noted.

Mould Dimension
Height of the mould : 170mm
Diameter of the mould : 150mm

Natural river sand was filled over the entire mould, and then the mould with sand weight was noted. The sand filled mould was fixed on the vibrating table. The table vibration was rapidly increased within 8 minutes. The effect of the settlement was noted using scale. Then the geomaterial (aggregate) was filled in the centre of the mould using a PVC pipe. Geomaterial was placed in three layers and each layer was compacted well with 24 blows. The mould with the sample was taken to the vibratory table and the same procedure is repeated which was carried before. Settlement was noted at the end. As a result, the settlement of sand stabilized using a geomaterial column will be less than before without the geomaterial. The pictures of experimental setup, mould filled with natural river sand and mould filled with natural river sand and geomaterial are shown in figure 1 through figure 4.

5. Results and discussions

Two tests were carried out i.e, only natural river sand and river sand stabilized with geomaterial. The diameter ratio was calculated (ratio of diameter of the geomaterial column to the diameter of the mould) as 1/5, 1/3, 7/15. For every diameter ratio value, the above mentioned two tests were done. Four trials have been done for each tests. From these tests the diameter ratio value of 7/15 was found to have less settlement when compared with the other two values.

Figure 1. Mould with natural river sand before vibration.

Figure 2. Mould with natural river sand after vibration.

Figure 3. Mould with natural river and stabilized with geomaterial before vibration.

Figure 4. Mould with natural river and stabilized with geomaterial after vibration.

In each trials, the settlement of material was noted after vibration. The value of settlement without geomaterial after vibration and the average value of settlement is given in Table.4.1. For this average settlement value, relative density was calculated which is also given in Table.4.1. The value of settlement with geomaterial after vibration and the average value of settlement is given in Table.4.2. For this average settlement value relative density was calculated which is also given in Table.4.2.

Table 1. Settlement of only river sand after vibration

Test no	Settlement after the vibration (mm)	Average settlement after the vibration (mm)	Relative density value
1.	32, 34, 31, 32	32.25	γ min=1.549g/cc γ max=1.911g/cc
2.	30, 29, 30, 31	30	γ min=1.583g/cc γ max=1.922g/cc

Table 2. Settlement of river sand stabilized with geomaterial after vibration

Test no	Settlement after the vibration (mm)	Average settlement after the vibration (mm)	Relative density value
1.	31,30,32,30	30.75	γ min =1.549g/cc γ max=1.891g/cc
2	28,27,25,27	26.75	γ min=1.583g/cc γ max=1.879g/cc

The loose sand condition is maintained at the start of the test by first making the aggregate column at the center of the mould by filling the PVC pipe with aggregate. Then the sand is filled in loose condition surrounding the pipe and pipe is removed. This prevents the densification before the start of the experiment. The settlement value after stabilization is only 26.75mm when compared to 30mm before stabilization. The results shows that the use of aggregate column within the river sand has reduced the settlement of loose sand in the mould. Thus the aggregate column due to its coarser nature and particle arrangement, tends to dissipate the vibrational effect being passed on to the surrounding soil. Hence the construction of aggregate/geomaterial column reduces vibration induced settlement.

6. Conclusions

Hereby we conclude that the better way to reduce settlement in loose sandy soil during and after tunnel construction is to stabilize with geomaterial column or wall. This will reduce settlement to greater extent as compared to tunnel provided only in loose sand deposit.

We suggest to use of aggregate to soil area ratio of 7/15 to get better reduction of settlement during and after tunneling. Thus, stabilizing the loose sand deposit with aggregate columns along the length of the tunnel or stabilizing the area between the tunnelling path and the building line will reduces the vibrational effect of the tunnelling operation and safeguards the buildings

References

[1] Boominathan, A., Subhadeep Banerjee and Priya Beena Sudevan 2019 "Numerical analysis on mitigation of liquefaction induced uplift of shallow tunnels", *Proc. of the XVII ECSMGE*, ISSMGE, Iceland.

[2] Lixiang Xie, Wenbo Lu, Jincai Gu, and Gaohui Wang 2018 "Excavation method of reducing blasting vibration in complicated geological conditions." *Shock and Vibration*, **2018(1)**, 1-

12.

[3] Huang, Jiankun and Shi, Zhifei 2015 "Vibration reduction of plane waves using periodic in-filled pile barriers." *Journal of Geotechnical and Geoenvironmental Engineering*, **141**, 6, 1-12.

[4] Deckner, F. 2017 "Vibration transfer process during vibratory sheet pile driving: from source to soil." PhD dissertation, KTH Royal Institute of Technology, Stockholm.

[5] Liu, Jinglei, Guishuai Feng, Jian Zhang, Xiaoyu Zhao, Chuanqing Yu, and Min Zhao. "Vibration isolation mechanism of concrete piles for Rayleigh waves on sand foundations." *Shock and Vibration*, **2018**(1), 1-13.

Influence of Addition of Glass Fibre on Concrete

NPannirselvam[1], SManivel[2]
[1]Associate Professor, Department of Civil Engineering,
[2]Assistant Professor, Department of Civil Engineering,
SRM Institute of Science and Technology, Potheri, Kattankulathur - 603203,
Chengalpattu District, Tamilnadu, India.
Corresponding author Email: [1]pannirsn@srmist.edu.in

Abstract. Fibres used to develop high performance concrete. Fibres such as glass, carbon, polypropylene and aramid fibres need to be improved in view of global sustainable development. Long term durability is achieved with the development of a high alkaline resistance fibre dispersion. This system was called reinforced concrete with glass fibre alkali resistance. To decide compressive and split rigidity, modulus of elasticity, modulus of crack and quick chloride penetrability tests for concrete grade M20 and M30 with and without fluctuating level of glass strands. The specimens were cast and tested at 28 days. The percentage of addition of glass fibres marginally increased the strength of compressive, flexural, split tensile for M20 and M30 grades of concrete when specimens tested at 28 days. A reduction in bleeding and permeability is also observed.

Key words: Concrete, Glass Fibre, Strength.

1. Introduction

Concrete was durable material and requires a little or no maintenance. As per IS: 456-2000, the durability aspects of concrete, has better experiences with durability of concrete structures [9,17,18]. Glass fibres are a recent introduction to the production of concrete fibres [7,11]. During the setting phase, concrete mixtures tend to shrink plastic and lead to cracking [1,6]. Adding small quantities of fibres can effectively eliminate cracking plastic shrinkage in the early age.

2. Objectives

1. To design for M20 and M30 grade of concrete.
2. To evaluate the mechanical properties of concrete with and without glass fibre in varying percentage.
3. To estimate the durability properties of concrete with and without glass fibre in varying percentages.

3. Experimental Investigations

Investigations were carried out on 48 cubes, 16 cylinders, 8 prisms and 8 RCPT mould with and without glass fibres of varying percentages. Trials on cast specimens were carried out to determine the strength of compressive, split tensile, flexural, modulus of elasticity and rupture and durability properties of both reference concrete and glass fibre concrete mixes.

4. Materials

4.1 Cement

In the investigation, ordinary Portland cement (53 grade) available on the local standard brand market was used. The details are given in Table 1.

Table 1. Physical properties

Sl. No.	Description	Values
1	Consistency	32%
2	Initial setting time	115 minutes
3	Final setting time	489 minutes
4	Specific gravity	3.49
5	Fineness of cement	4%
6	Compressive strength at 28 days	58.86 N/mm^2

4.2 Coarse aggregate

The coarse aggregate tested for its various properties and details are tabulated in Table 2 and 3.

Table 2. Sieve analysis

Sl. No.	Sieve Size(mm)	Percentage of Passing
1	40	100.00
2	20	82.72
3	12.5	15.22
4	10	2.94
5	4.75	0.22
6	Less than 4.75	0

Table 3. Specific gravity and bulk densities for coarse aggregates

Sl.No.	Size of Coarse Aggregate	Specific Gravity	Bulk Density (Kg/m^3)	
			Loose	Rodded
1	20 mm	2.65	1622	1781

4.3 Fine aggregate

The fine aggregate tested for its physical properties and results are shown in Table 4 and 5.

Table 4. Sieve analysis.

Sl. No.	Sieve size (mm)	Cumulative percentage of passing	Remarks
1	10.0	100.00	
2	4.75	94.90	
3	2.36	93.30	The tested sand belongs to Zone – II category.
4	1.18	71.90	
5	0.600	40.80	
6	0.300	8.50	
7	0.150	1.30	

Table 5. Specific gravity and bulk densities of supplied fine aggregate.

Sl. No.	Fineness modulus	Specific gravity	Bulk density (Kg/m^3)	
			Loose	Rodded
1	2.76	2.62	1598	1732

4.4 Glass fibres

Cem-FIL® Anti–Crack High Dispersion used for the experimental study and their properties were furnished by the manufacture. The samples of Cem-FIL® Anti–Crack High Dispersion are presented in Figs. 1 and 2.

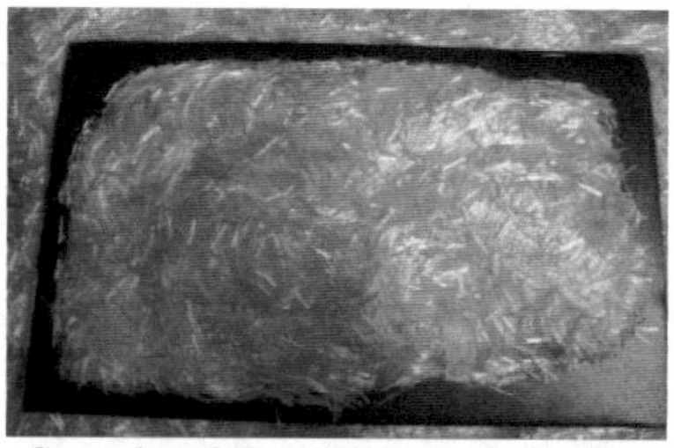

Figure 1. Sample of Cem-FIL Anti-Crack HD Fibre.

Figure 2. Microscopic Structure of Cem-FIL Anti-Crack HD Fibre.

4.5 Water

Water used for making concrete from the laboratory used for drinking.

4.6 Concrete Mix Design

Concrete design for grades M20 and M30 as per IS: 10262 – 2019 and the proportions are presented in Table 6.

Table 6. Mix Design.

Grade of Concrete	Cement (kg)	Fine Aggregate (kg)	Coarse Aggregate (kg)	Water (Lts)	W/C Ratio	Glass Fibres
M 20	320	731	1119	176	0.55	0.03, 0.06 & 0.10% by concrete volume
M 30	350	686	1137	175	0.50	

5. Casting of test specimens

The present experimental study includes casting and testing of specimens for compressive, split tensile, flexural strength and rapid chloride permeability test are performed. The specimens are casted using concrete design mixes.

5.1. Mixing

Pan mixing is adopted throughout the experimental work (Figure. 3) for casting of the specimens.

Figure 3. Mixing of Concrete (Pan Mixer).

5.2. Cast specimens

A standard moulds of size 100x100x100 mm cubes, 150 mm diameter x 300 mm height and 100mm diameter x 200mm height cylinders and 100x100x500 mm beam moulds are used. The specimen after casting is shown in Figure.4.

Figure 4. Specimens after Casting.

5.3. Compaction of concrete

In the present investigation, the table vibrator is used for compacting the concrete as shown in Figure5. Compaction of concrete by vibration makes the concrete better quality, higher strength with given cement content with less mixing water.

Figure 5. Specimen Induced for Vibration.

5.4. Curing of specimen

The cast specimens are removed from the mould which left for a day at room temperature following curing. The samples are submerged in the curing tank for 28 days. After curing the specimens dried in sunlight and tested to evaluate its hardened strength.

6. Testing of specimens

The cast specimens are tested after it hardens.

6.1. Compressive strength

The test was performed on a 2000kN hydraulically operated digital CTM (Compression Testing Machine) and represented in Figure 6.

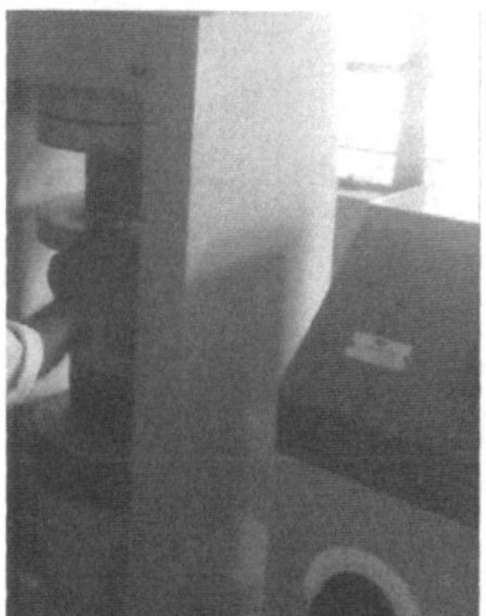

Figure 6. Evaluating the Compressive Strength.

6.2. Split tensile strength

The test is performed horizontally on hardened specimen between a CTM's loading surfaces and the load is applied along the vertical diameter until the cylinder fails. The details of the setup are shown in Figure 7.

Figure 7. Evaluating the Split Tensile Strength.

6.3. Flexural strength

Two point loading is applied to estimate the strength of concrete by conducting flexural test. The details of the specimen are presented in Figure 8.

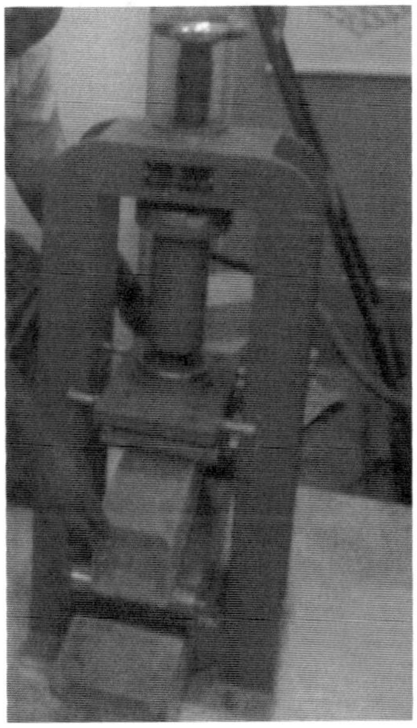

Figure 8. Computing Flexural Strength of Beam.

6.4. Secant modulus of elasticity

Secant modulus elasticity of cement is 33 percent ultimate strength as shown in Figure 9.

Figure 9. Computing Modulus of Elasticity.

6.5. Testing for Rapid Chloride Permeability Test

The RCPT specimens (cylinder diameter of 150mm and height of 200mm) are cast and cured for 28 days and loaded in the cells as shown in Figure. 10.

Figure 10. Rapid Chloride Permeability Test with Specimens.

7. Results and discussion

The information acquired from exploratory examinations on the modulus of elasticity, modulus of crack, compressive strength, split tensile strength and fast chloride penetrability test results are introduced in the accompanying areas. The expansion of glass filaments and their conduct of the cast examples are additionally exhibited.

7.1. Compressive Strength

The compressive strength for M20 and M30 grades with different percentage of fibres for 7 and 28 days are presented in Table 7.

Table 7. Compressive Strength for different mix proportions.

		Compressive Strength N/mm^2									
S.No	Grade	0%		0.03%		0.06%		0.10%		0.15%	
		7days	28 days	7 days	28 days	7 days	28 days	7 days	28 days	7 days	28 days
1	M20	16.68	27.37	17.75	34.84	18.47	36.28	19.21	38.54	17.31	35.45
2	M30	23.75	37.39	26.67	45.22	27.67	47.62	29.05	49.05	26.05	44.50

7.2. Split Tensile Strength

M20 and M30 grade's split tensile strength is determined with various level of fibres at 28 days are enlisted in Table 8.

Table 8. Split Tensile Strength for various percentage of Glass Fibre.

S.No	Grade	0%	0.03%	0.06%	0.10%	0.15%
1	M20	3.61	4.05	4.33	4.45	4.15
2	M30	4.23	4.68	4.89	5.12	4.56

7.3. Modulus of Elasticity

M20 and M30 grade's modulus of elasticity is determined with various level of fibres at 28 days are enlisted in Table 9.

Table 9. Modulus of Elasticity for various percentage of Glass Fibre.

S.No	Grade	0%	0.03%	0.06%	0.10%	0.15%
1	M20	22478	24356	23486	23260	23140
2	M30	28643	28978	27978	27618	27149

7.4. Modulus of Rupture of Beams

The modulus of rupture of beams for M20 and M30 grades with different percentage of fibres at 28 days are presented in Table 10.

Table 10. Modulus of Rupture of Beams for various percentage of Glass Fibre.

S.No	Grade	0%	0.03%	0.06%	0.10%	0.15%
1	M20	3.52	3.96	4.18	4.29	4.12
2	M30	4.12	4.57	4.96	4.98	4.16

7.5. Rapid Chloride Permeability Test

RCPT for M20 and M30 grades with different percentage of fibres at 28 days are presented in Table 12.

Table 11. RCPT for various percentage of Glass Fibre.

S.No	Grade	0%	0.03%	0.06%	0.10%	0.15%
1	M20	3950	3343	3472	3600	3750
2	M30	2962	2272	2343	2355	2487

8. Conclusions

1. The rate increment in compressive strength of various glass fibre concrete proportion is seen from 20 to 40 percent contrasted with 28 days compressive quality.
2. The rate increment in flexural and split tensile strength of glass fibre solid blends of various evaluations is seen from 15 to 25 percent contrasted with 28 days.
3. Reduced bleeding is observed by adding glass fibres which improves its homogeneity and reduces cracking.

4. For higher concrete evaluations, the penetrability of glass fibre strengthened chloride concrete shows less chloride porousness.
5. Generally, glass fibres reduce cracks that cause vacuum interconnection to be minimal.
6. The penetrability of cement with the addition 0.03% of glass fibres in M20 and M30 grades diminishes at 28 days by 15.37 and 23.29%.

9. Reference

[1] Lim C C, Gowripalan N and Sirivivatnanon V 2000Microcracking and Chloride Permeability of Concrete under Uniaxial Compression *Cem. Concr. Compos.* **22**353-60.

[2] Caijun Shi 2004 Effect of Mixing Proportions of Concrete on its Electrical Conductivity and the Rapid Chloride Permeability Test (ASTM C1202 or ASSHTO T277) Results Cement and Concrete Research **34** 537-45.

[3] Rolf Feldman, Luiz RPrudencio Jr U and Gordon Chan 1999 Rapid Chloride Permeability Test on Blended Cement and other Concretes: Correlations between Charge, Initial Current and Conductivity*Constr. Build. Mater.***13** 149-54.

[4] ErhanGuneyisi, Mehmet Gesoglu, TuranOzturan and Erdogan Ozbay2009 Estimation of Chloride Permeability of Concretes by Empirical Modeling: Considering Effects of Cement Type Curing Condition and Age*Constr. Build. Mater.***23** 469–81.

[5] Chindaprasirt P, Rukzon S and SirivivatnanonV 2008 Resistance to Chloride Penetration of Blended Portland Cement Mortar Containing Palm Oil Fuel Ash Rice Husk Ash and Fly Ash*Constr. Build. Mater.***22** 932–38.

[6] Barluenga G and Hernández-Olivares F 2007 Cracking Control of Concretes Modified with Short AR-Glass Fibers at Early Age Experimental Results on Standard Concrete and SCC Cement and Concrete Research **37** 1624-38.

[7] Asokan P, Osmani M and Price A D F 2009 Assessing the Recycling Potential of Glass Fibre Reinforced Plastic Waste in Concrete and Cement composites*J. Clean. Prod.***17** 821–9.

[8] Perez-Pena M and Mobasher B 1994Mechanical Properties of Fiber Reinforced Lightweight Concrete CompositesCement and Concrete Research **24** 1121-32.

[9] Purnell P and Beddows J 2005 Durability and Simulated Ageing of New Matrix Glass Fibre Reinforced Concrete*Cem. Concr. Compos.* **27** 875–84.

[10] Faiz A Mirza and ParvizSoroushian 2002 Effects of Alkali-Resistant Glass Fiber Reinforcement on Crack and Temperature Resistance of Lightweight *Cem. Concr. Compos.***24** 223-7.

[11] Purnell P, Short N R, Page C L and Majumdar A J 2000 Microstructural Observations in New Matrix Glass Fibre Reinforced Cement Cement and Concrete Research 30 1747-53.

[12] OrmelleseM, Berra M, Bolzoni F andPastoreT 2006 Corrosion Inhibitors for Chlorides Induced Corrosion in Reinforced Concrete StructuresCement and Concrete Research**36** 536 – 47.

[13] Dale P Bentz2007 A Virtual Rapid Chloride Permeability Test*Cem. Concr. Compos.***29** 723-31.

[14] Julio-Betancourt G A and Hooton R D 2004 Study of the Joule Effect on Rapid Chloride Permeability Values and Evaluation of Related Electrical Properties of Concretes Cement and Concrete Research **34** 1007-15.

[15] Wee T H,Arvind K Suryavanshi and San San Tin 1999 Influence of Aggregate Fraction in the Mix on the Reliability of the Rapid Chloride Permeability Test*Cem. Concr. Compos.***21** 59-72.

[16] Yuwaraj M Ghugal and Santosh B. Deshmukh2006 Performance of Alkali-resistant Glass Fiber Reinforced Concrete Journal of Reinforced Plastics and Composites**25**617-30.

[17] Chandramouli K, Srinivasa Rao P, Pannirselvam N and SeshadriSekhar T 2010 Long Term Durability Studies on Glass Fibre Reinforced Concrete Using Rapid Chloride Permeability Test International Journal of Applied Engineering Research **5**557–64.

[18] Chandramouli K, Srinivasa Rao P, Pannirselvam N and SeshadriSekhar T 2010 Study on Strength and Durability Characteristics of Glass Fibre Concrete, International Journal of Mechanics and Solids**5**15-26.

Experimental Investigation of Recycled Coarse Aggregates on Partial Replacement of Cement with Silica Fume

SManivel[1], NPannirselvam[2], JRajprasad[3], T.SHemanarayanan.[4]
[1]Assistant Professor, Department of Civil Engineering,
[2]Assosciate Professor, Department of Civil Engineering,
[3]Assistant Professor (Sr.G), Department of Civil Engineering,
[4]PG Student, Department of Civil Engineering,
SRM Institute of Science and Technology, Potheri, Kattankulathur 603203, Chengalpattu District, Tamilnadu, India.
Corresponding Author Email: [2]pannirsn@srmist.edu.in

Abstract. Coarse aggregate utilized majorly in concrete to reach the ultimate quantity in larger proportion. The structural and non-structural elements in the construction utilize majorly the concrete. Due to the mass availability of concrete during the demolition process and hence absorbing the same material to replace the coarse aggregate with recycled aggregate is the emerging field and it is sustainable. The experimental work involving the replacement of coarse aggregates with recycled aggregates and cement with silica fumes to improve the properties of concrete and eliminate the difficulties in construction field paves way for the further researches. The experimental study concentrates in the strength properties of the concrete with recycled aggregates and silica fume and compared with the conventional concrete.

Keywords: Compressive strength, Recycled coarse aggregate, Silica fume.

1. Introduction

Concrete is a combination of cement, fine grained particles and coarse aggregates binded with water and hardens due to many external factors. Several structural elements such as slabs, columns, beams, footings and lintels etc. are built using concrete. In recent years, many researches have been done in the replacement of components of concrete with the chemicals, fibres, agro wastes, construction and demolition wastes.

Concrete is brittle and possess good compression but weak in tension. The presence of cement causes cracking of beams, columns and cannot be operated if the moisture of soil is above 10%. The disposal of cement causes emission of carbon dioxide, which affects the environment and causes pollution. Hence the cement can be partially replaced with silica fume[1-3]. Silica fume possess higher fineness texture and has extreme amorphous silica content, so this silica based fume can act as a high reactive factor because of its pozzolanic reaction [8, 12]. The main advantage of the silica fume on addition to the concrete ensures durability enhances flexural and compressive strength [4-6].

In recent years, the usage of recycled coarse aggregate (RC) have an effective potential in production of concrete is more considerate, as it is environmentally concerned and conserves the natural resources. Construction and demolition activities accounted for 375million and from the demolition of bridge sand roads produces huge amount of debris. Disposal of this aggregate creates an issue on waste disposal and increase in cost of waste treatment prior to disposal. Recycled concrete aggregates from demolished building or structures used in production of concrete [9-11].

2. Objectives

The important objectives of the research includes

- To optimize the addition of percentage of recycled aggregate.
- To optimize the addition of percentage of silica fume.
- To observe the compressive strength of addition of 10, 20 and 30 % of recycled aggregate along with 10, 15 and 20% of silica fume.

3. Materials

3.1 Silica Fume

The particle size of the silica fume adopted for the experimental work is 150 nano meter and shown in Figure 1.The specific gravity of coarse aggregate is 2.24.

Figure 1. Silica fume.

3.2 Recycled Coarse Aggregate

Recycled aggregates are the aggregates which are the resultant product from the construction demolition wastes. It is adopted in the mixing of concrete by partially substituting the natural aggregates shown in Figure 2.

Figure 2. Recycled coarse aggregate.

3.3 M sand

M sand also called as quarry powders, quarry minerals, quarry flour, soil demineralization and mineral fines having finely crushed rock, obtained by natural or mechanical means, which contains minerals and trace elements largely used in organic farming cultures are shown in Figure 3. The test results are presented in Table 1.

Table 1. Properties of M sand

Property of material	Results
Specific gravity	2.47
Fineness modulus	3.98
Sieve analysis	Zone II

Figure 3.M sand.

3.4 Coarse aggregate

A property of coarse aggregate consists of porosity, density, grading, shape and surface texture which evaluates the characteristics of concrete mixtures. The coarse aggregates used in the experimental work is shown in Figure4. The specific gravity of coarse aggregate is 2.74.

Figure 4. Coarse aggregate

3.5 Cement

The cement used in the research work from the locally available market of 53 grade are shown in Figure5. The physical properties of cement presented in Table 2.

Table 2. Properties of cement

Properties of material	Results
Cement consistency	31%
Cement fineness	5%
Initial setting time	39 minutes
Final setting time	489 minutes
Specific gravity	3.14

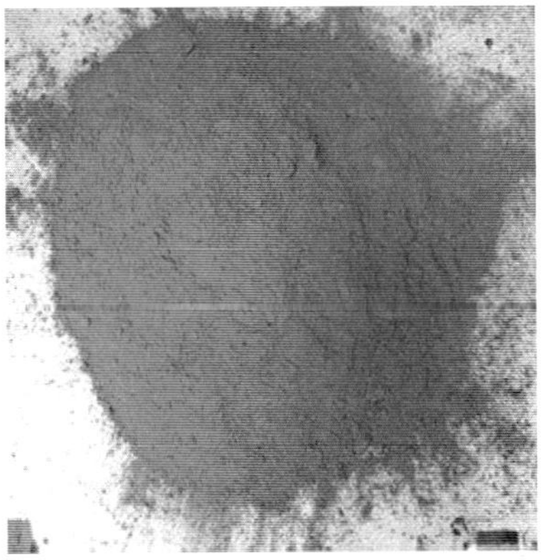

Figure 5. Cement.

3.6 Water

Water was one of the most environmental elements in construction it increase the engineering properties of fine-grained particles.

4. Methodology

The cubes were casted and checked for compressive strength and the flexural properties. These tests are accomplished to examine the characteristics of toughness of concrete without admixtures and on addition of these additives, the rate of strength and enhancement of durability can be found. The design mix were calculated for M40 conventional concrete and concrete with silica fumes of proportions 10, 15 and 20%. A concrete mix proportions of 1: 1.5: 2.65 with water cement ratio of 0.40 under laboratory conditions.

5. Results and discussion

The compressive, split tensile and flexural strength of the specimen after curing are tabulated in Table 3.

Table 3. Compressive, Split Tensile and Flexural Strength.

S.No.	Sample	Compressive strength, N/mm^2		Split tensile strength, N/mm^2	Flexural strength, N/mm^2
		7 Days	28 Days	28 Days	28 Days
1	Control specimen	27.33	38.16	3.76	5.48
2	10% silica 10% RC	24.10	34.90	3.48	5.39
3	10% silica 20% RC	24.86	30.16	3.02	5.22
4	10% silica 30% RC	25.38	29.74	2.89	4.97
5	15% silica 10% RC	25.47	33.27	3.33	5.24
6	15% silica 20% RC	26.92	36.77	3.68	5.88
7	15% silica 30% RC	26.28	31.14	3.24	5.42
8	20% silica 10% RC	25.80	32.86	3.29	5.10
9	20% silica 20% RC	26.44	34.13	3.43	5.32
10	20% silica 30% RC	24.69	29.75	2.89	4.72

6. Conclusions

The addition of percentage of recycled aggregate at 20% with addition of 15% silica fume are optimized. The compressive strength of addition of 20 % of recycled aggregate along with 15% of silica fume at 28 days is 36.77 N/mm^2. Further addition leads to reduction of compressive strength. The split tensile strength of addition of 20% of recycled aggregate along with 15% of silica fume at 28 days is 3.68 N/mm^2. The flexural strength of addition of 20% of recycled aggregate along with 15% of silica fume at 28 days is 5.88 N/mm^2.

7. References

[1] Belen G F, Fernando M A, Isabel M Land Javier E L 2009 Structural shear behavior of recycled concrete with silica fume*Constr. Build. Mater.* **23** 3406–10.

[2] Chakradhara Rao M 2011 Recycled coarse aggregate and its use in concrete *ICI journal* 27- 40.

[3] Elhakam A A, Mohamed A E M and Awad E 2012 Influence of self healing, mixing method and adding silica fume on mechanical properties of recycled aggregates concrete, *Constr. Build. Mater.* **35**421–7.

[4] Evangelista L and De Brito J 2007 Mechanical behavior of concrete made with fine recycled concrete aggregates *Cem. Concr. Compos.***29**397–401.

[5] Huda S B and Alam M S 2014 Mechanical behavior of three generations of 100% repeated recycled coarse aggregate concrete*Constr. Build. Mater.* **65**574–82.

[6] Kou S and Poon C 2008 Long-term mechanical and durability properties of recycled aggregate concrete prepared with the incorporation of fly ash*Cem. Concr. Compos.***37**12-19.

[7] Pinghua Zhu,YaliHao, Hui Liu, Da Wei, Shaofeng Liu and Lei Gu2019 Durability evaluation of three generations of 100% repeatedly recycled coarse aggregate concrete*Constr.*

Build. Mater. **210**442–450.
[8] Priyanka andMusaibMehraj Bhat 2015 Study the effect of fly ash, silica fume and recycled aggregate on the compressive strength of concrete aggregate on the compressive strength of concrete, *International journal of research in engineering and advanced technology*, **3** 71-7.
[9] Rao A, Jha K N, andMisra S 2007 Use of aggregates from recycled construction and demolition waste in concrete*Resources, conservation and recycling***50**81–91.
[10] Ryu J S 2002 An experimental study on the effect of recycled aggregate on concrete properties*Magazine of concrete research***54** 7–12.
[11] TusharRSonawane and Sunil SPimplikar 2013 Use of recycled aggregate concrete*Journal of mechanical and civil engineering*52-9.
[12] OzgurCakır and Omer OzkanSofyanli 2015Influence of silica fume on mechanical and physical properties of recycled aggregate concrete*Housing and building national research center***11**157-66.

A Study on Polymeric Fibre Reinforced Stabilized Mud Blocks

N.P. Vignesh [1], N. Arunachelam [2], K. Mahendran [3] and B. Dinesh Kumar [4]

[1,2,3] Centre for Rural Technology, The Gandhigram Rural Institute Deemed to be University, Dindigul, India.
[4] Department of Civil Engineering, Kongu Polytechnique College, Perundurai, India.

np.vignesh@gmail.com

Abstract. It is a known fact that mud blocks casted out of black cotton soil have poor damage resistance due to clayey, brittle and weak nature of the black cotton soil. In order to improve the strength of the cotton soil based mud blocks , additives such as polypropylene filament , polypropylene flakes and poly styrene fibres have been used in addition to a natural binder namely Terminalia chebula. By varying the percentage of the additives and as also soil, sand and lime various fibre reinforced blocks are made. The as-prepared reinforced mud blocks on analysis were found to possess a maximum strength of 13.75MPa when polypropylene was added as a additive. The study revealed that the addition of polymeric additives with the use of Terminalia chebula resulted in reinforcement of the stabilised mud blocks with enhanced strength and durability.

Keywords: Artificial Fibres, Lime Stabilization, Terminalia chebula, Fibre Reinforcement, Compressed Stabilized Earth blocks, Stabilized Mud Blocks.

1. Introduction

Shelter for all is a major and important goal of all nations around the world. Despite best efforts, still India needs 27 million units of housing. There has been not only a huge cost escalation of materials but also a high demand of construction materials. Housing has gone out of the reach of middle class and economically weaker sections of the society. Still there is a need to provide housing to a large population in India. In the absence of appropriate strategiesfor the housing, the shortage is likely to shoot up to 38 million units in 2030. It was estimated that around 30% of world's population are still having earthen structures. In view of the abundant availability, low cost and eco friendly nature, earth still used as extensively in building constructions. In fact, earthen masonry encourages the green building movement. Further it reduces the use of manufactured materials and consumes low energy. If the soil used for building constructions contains clay materials structural problems result. The advancements in the field of Material Science and Engineering have not developed earth as a good building material. Several researches have embarked on the use of stabilized mud blocks as building materials with improved strength and durability. There have been many attempts addressed such problems like low tensile strength, brittleness and deterioration on exposure to water [1-6]. The present study is on the production of mud blocks using black cotton soil with polymeric fibres on additives and *Terminalia chebula* as natural binder. The results of the study are presented in this communication.

2. Materials and Methods

2.1 Black Cotton Soil

The black cotton soil used for preparing mud blocks for this study is collected from Athoor block of Dindigul District Tamilnadu. The soil sample was collected about 2 m below the ground level. The physical properties are given in table 1, while the percentage chemical composition of the black cotton soil is given in table 2. Similarly the percentage of lime treated black cotton soil is given in the table 3.

2.2 Lime

The lime used for our study was obtained locally. The chemical composition of lime is given in table 4.

2.3 Terminalia chebula

Locally collected plants of *Terminalia chebula* were cut into small pieces and soaked in to water at a ratio of 1:15 with the solution brewing for nearly 24 hr prior to use [7]. The major phytoconstituents present in *Terminalia chebula* are hydrolysable tannins, poly phenols, gallic acid, chebulagic acid and chebulinic acid [8].

2.4 Sand

River sand passing through 2 mm IS sieve and retain over 425 micron sieve was used for this study. The specific gravity used was 2.67.

2.5 Fibre

In order to overcome the drawback of the lack of strength, polymeric fibres were added. To find out the effect of the morphology of the polymer on the strength and durability of the mud block, polypropylene flakes and polypropylene mono filaments were added. In order to study the effect of nature of the polymer over the properties of the mud blocks polystyrene fibres were also added and tested. These polymeric materials were obtained commercially from the vendors and used as such. The properties of fibres are given in the table 5.

3. Experimental Methods

Perusal of the properties of the black cotton soil shows that the plasticity index of the soil is fairly high. It is pertinent to observed that Walker, P.J., et.al (1995) had observed that soil with plasticity index higher than 25 needs only machine compaction. This higher plasticity index is due to higher clay percentage in the soil (table 1).

Table 1. Properties of Black Cotton Soil

Colour	Dark brown
Specific gravity	02.03
Silt (%)	30.00
Clay (%)	70.00
Swelling index (%)	100.00
Liquid limit (%)	55.00
Plastic limit (%)	07.70
Plasticity index	47.30
Dry density(g/cc)	01.49
Optimum moisture content (%)	20.00

IOP Conf. Series: Materials Science and Engineering 1026 (2021) 012010 doi:10.1088/1757-899X/1026/1/012010

Table 2. % of Chemical Composition of Black Cotton Soil

Chemical	SiO_2	Al_2O_3	Fe_2O_3	CaO	MgO	P_2O_5	K_2O	SO_3
%	62.4	16.1	11.6	5.1	1.7	0.4	1.8	0.5

Table 3. Properties of Lime Treated Soil

Optimum moisture content (%)	22.00
Maximum dry density (g/cc)	0.55
Liquid limit (%)	46.20
Plastic limit (%)	10.50
Plasticity index	35.70

Table 4. Chemical Composition of Lime

Elements	% Existing
O	28.92
Ca	55.72
C	10.24
Si	02.76
Mg	0.88

Table 5. Properties of The Fibres Used

Material	Polystyrene	Polypropylene
Density(g/cm ^{3})	1.34	0.91
Tensile strength (MPa)	600	≥500
Elongation rate	≥15%	≥15%
Acid resistance	Fine	Fine
Length of fiber(mm)	10	10
Diameter of fiber(µm)	25-30	38
Melting point (°C)	240-260	160 -180
Burning point (°C)	>500	>360
Color	White	White
Elongation rate	≥15%	≥15%

The percentage chemical composition of the soil reveals low calcium content and hence calls for lime supplementation in the mud block casting process. The treatment of lime to the black cotton soil has brought down the plasticity index from 47.3 to 35.7 (table 3). Both the polymeric materials used in this study are thermoplastics in nature and polystyrene has a greater tensile strength than polypropylene (table 5).

3.1. Mixing, Moulding and Curing

Various mixing procedures have been highlighted for casting and curing of the stabilized mud blocks [9 -11]. The black cotton soil was dried, the lumps were broken and the soil was sieved using 4.75 mm to remove organic matters and waste particles. The various mix proportions used for the making of stabilized mud blocks are shown in table 6.

Table 6. Mix Proportions Evolved in the Making of Polymeric Reinforced Mud Blocks

Mix designation	Composition in %					
	Soil	Sand	Lime	Polypropylene Flaked	Polypropylene Monofilament	Polystyrene
S0	90.0	5	5	-	-	-
S1	89.9	5	5	0.1	-	-
S2	89.8	5	5	0.2	-	-
S3	89.7	5	5	0.3	-	-
S4	89.6	5	5	0.4	-	-
S5	89.5	5	5	0.5	-	-
S6	89.9	5	5	-	0.1	-
S7	89.8	5	5	-	0.2	-
S8	89.7	5	5	-	0.3	-
S9	89.6	5	5	-	0.4	-
S10	89.5	5	5	-	0.5	-
S11	89.9	5	5	-	-	0.1
S12	89.8	5	5	-	-	0.2
S13	89.7	5	5	-	-	0.3
S14	89.6	5	5	-	-	0.4
S15	89.5	5	5	-	-	0.5

During this mixing, the fibres were added by hand in stages, to form a homogeneous soil matrix. Mixing was continued to ensure a uniform distribution of fibre throughout the soil matrix. The wet mixing was done by adding *Terminalia chebula* extract to soil-lime-fibre matrix. The percentage of *Terminalia chebula* extract was maintained as 20% common for all the mix proportions. The blocks were compacted immediately after the mixing and casted using Auroville Ramming Machine (AURAM). The constant moulding pressure of 15 MPa was given by a single acting ram of AURAM machine. The specification of the manual press was given in table 7. The blocks were then tested for compaction by pressing the penetrometer on the surface of the mud blocks. The depth of penetration indicated the quality of compaction. Adequate care was taken to see that the penetration was not more than 10%.

Table 7. Specifications of Manual Press Used

Arm of lever	Dimension of blocks	Number of blocks	Applied Force	Compression Mode	Compression category
1800mm	240 X 240 X 100mm	1	15 MPa	Static	Low

The blocks (S0 to S15) were kept under ambient curing for three days, and then the blocks were covered and cured by spraying water regularly for 28 days.

3.2 Testing

Initially the Unconfined Compressive Strength was found out and it was compared with the compressive strength of casted blocks tested at 7th, 14th, 21st and 28th day of curing in digital Compression Testing Machine of 3000 KN capacity. Also these mud blocks were subjected to water absorption and spray erosion tests after 28th day of curing to assess their performance in adverse weather conditions. The water absorption test was carried out based on IS: 3495 (Part II) 1976. The specimens were wetted for 24 hours and their weights are determined. The difference in weight of the mud blocks before and after immersion was used for the calculation of water absorption. The test setup for the modified spray erosion is shown in figure 1. The obtained results were tabulated and discussed below.

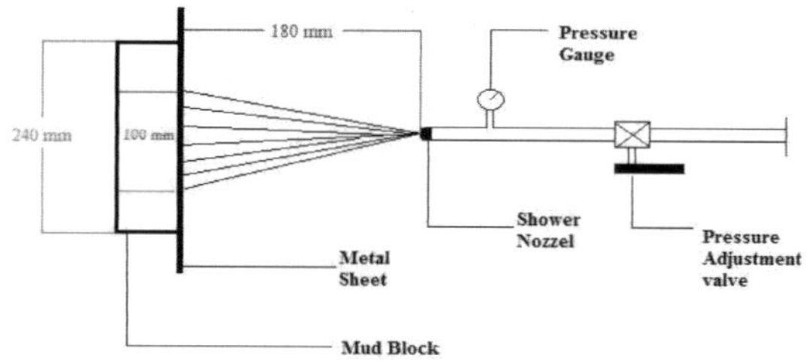

Figure 1. Working setup for modified spray erosion test

4. Results and Discussion

4.1 Unconfined Compressive Strength (UCS)

The utility of any material for construction purpose depends on the compression strength. In this study, the mud blocks were prepared with polymeric adjuvant with a view to reinforce the mud blocks. The UCS test was carried out to determine the strength of cohesive or bonded soils in an inexpensive and practical manner without the need for applying a confining stress, while maintaining soil bonding or cementation prior to shearing [12]. The increase in the compressive strength by the addition of polymeric fibres can be attributed to the reinforcement action of the polymeric fibres. This result endorses the earlier view that addition of polymeric fibres reduces shrinkage cracking; improve tensile strength, durability and ductility [9]. From the results of the study (table 8) it is evident that the maximum strength could be obtained in the case of S5, S10 and S15 among the same polymeric constituents. The results also point out that the beyond 0.5%, the increase in compressive strength is not much significant.

Table 8. Unconfined Compression Strength for All Mix

% of Fibers	Compressive Stress MPa		
	Polypropylene monofilament	Polypropylene flakes	Polystyrene
0.1	0.299	0.254	0.392
0.2	0.311	0.278	0.414
0.3	0.383	0.332	0.46
0.4	0.395	0.395	0.483
0.5	0.422	0.422	0.525

4.2 Compressive Strength Test

The compression strength was determined for all casted blocks (S0 to S15) at the 7th, 14th, 21st and 28th day of curing shown in the figure 2. The results show an apparent increase in the compression strength compared to the raw soil specimen. The matrix reinforced with the polymeric fibres at 0.5% recorded the highest compressive strength in the case of all three polymeric materials. The effect of fibre is so pronounced in polystyrene fibres. The highest compression strength of 13.75 MPa would be attained after 28 days of curing in the case of S15. Similar work was carried out by Hanifi Binici et.al and C.K.Subramania Prasad et. al [1, 6] by adopting plastic and natural fibres for the reinforcement of blocks with cement and lime as stabilizers. Comparison of results indicates that compressive strength registered an increase of nearly 2.5 times over the control. The higher strength attained in the present study may be due to the lime supplementation to the soil which resulted in good bondage between the soil particles with polymeric fibres acting as reinforcement materials [7].

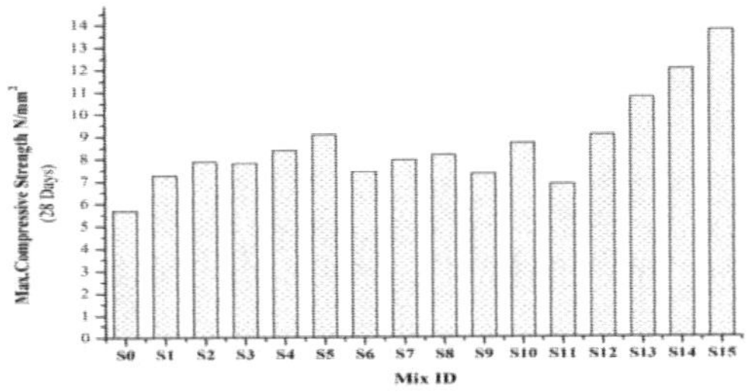

Figure 2. Compression strength of Mud Blocks at different mix proportions

4.3Water Absorption

The blocks were tested in accordance with the procedure laid down in IS:3495 (Part II) 1976, after immersion in cold water for 24 hours. As per code, the average water absorption shall not be more than 15% by its weight.

But in the case of S6 block, the water absorption was found to be nearly 29.2% which could be due to the poor compaction. Earlier workers have overcome this drawback by increasing the compaction pressure above 12.5 MPa and addition of sand was found to reduce the water absorption but did not affect the compression strength. But in the case of other samples, the water absorption was found to be much less. The addition of natural binder viz., *Terminalia chebula* could have reduced the voids, thus increasing the bonding between soil and sand which had resulted in less water absorption shown in figure 3 [14-17].

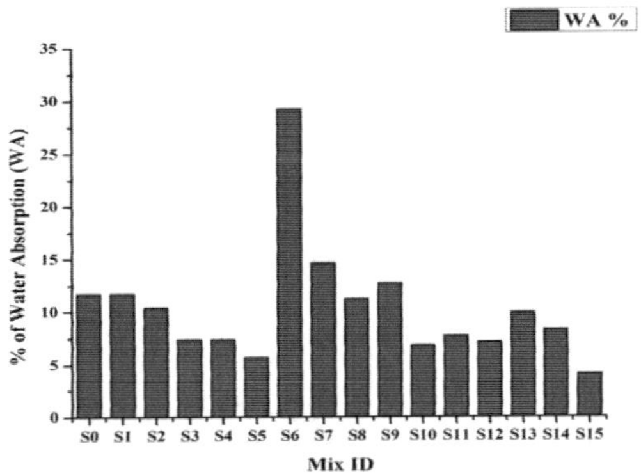

Figure 3. Water Absorption of Stabilized Mud Blocks

4.4 Spray Erosion

It is necessary to track the performance of the stabilized fibre reinforced mud blocks when it is exposed to adverse climatic conditions (rain, storm etc.,) in the field in order to assess the durability such a study may not feasible option because of the large duration of time and hence a laboratory level test setup had been developed with reference to IS: 1725-1982 specifications [19 & 20].
Here the erosion test had been conducted to S5, S10& S15 blocks which attained higher compressive strength in each kind of fibres used. The results of erosion test are given in the table 10. B.V.Venkataramareddy et.al [18] derived an equation for erosion rate is

$$ER = d/(P_r \, X \, t) \tag{1}$$

Where, D = avg. depth of the erosion
P_r = precipitation rate
t = duration of erosion in minutes

From the results, it is evident that the rate of erosion is minimum for polystyrene fibre reinforced mud blocks.

Table 9. Erosion Test Results of High Strength Mud Blocks

Mix Proportion	Time (minutes)	Depth of erosion in (mm)	Rate of Erosion in (mm/min)
S5	5	0	0.062
	10	0	
	15	0	
	30	0.75	
	60	0.80	
	90	1.00	
	120	1.25	
S10	5	0	0.059
	10	0	
	15	0.40	
	30	0.70	
	60	0.90	
	90	1	
	120	1.19	
S15	5	0	0.049
	10	0	
	15	0.10	
	30	0.35	
	60	0.50	
	90	0.85	
	120	1.00	

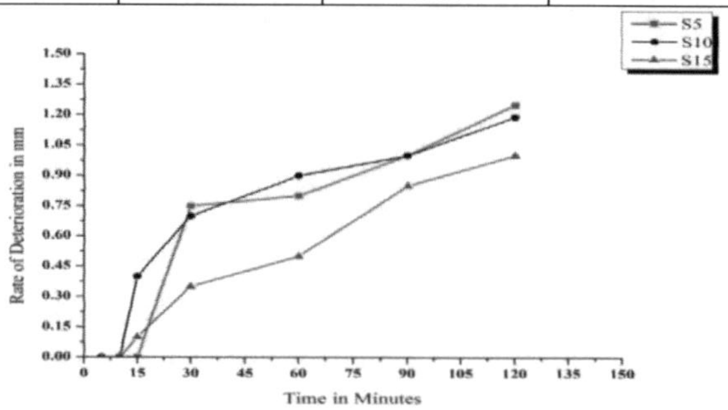

Figure 4. Spray Erosion pattern for High Strength Mud Blocks

The obtained values of the erosion are minimal and satisfies the IS codal recommendations which are shown in the figure 4. The erosion test reveals that with optimum stabilizer content of lime and *Terminalia chebula,* the fibres can be used without affecting the durability of the blocks [21]. The block possess minimal erosion rate and no extra care is needed to protect it [22]. The presence of *Terminalia chebula* extract acts as a natural water –proof coating on soil particles which may also be the reason for lower erosion since it acts as a barrier for the water entry.

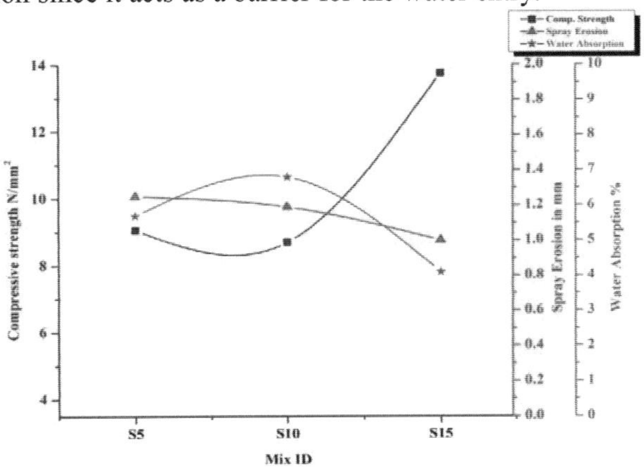

Figure 5. Comparison of Compressive strength, Spray Erosion and Water Absorption in various Stabilized Mud Blocks

Comparison of results obtained for various mud blocks casted with regard to such parameters like compression strength, water absorption and spray erosion clearly shows that there exists an inverse relationship between compression and spray erosion and water absorption.

4.5 Morphological Studies

Morphological studies were conducted for the samples which exhibited higher strength (S0, S5, S10 and S15) and the following inference could be arrived at.

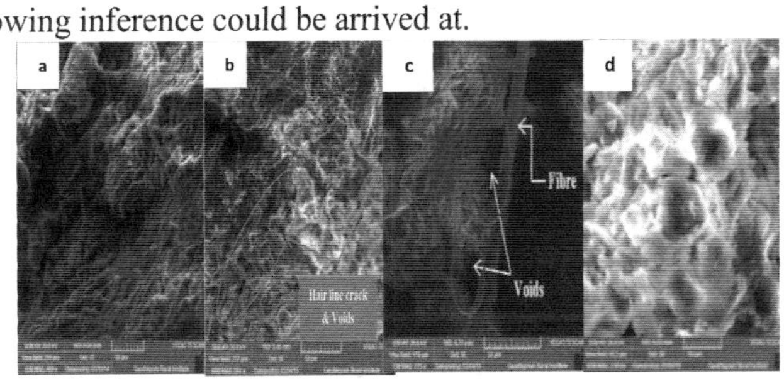

Figure 6. SEM Images of (a) S0, (b) S5, (c) S10 & (d) S15 samples

Figure 6 (a) shows that there was a tight packing and ingredients had bonded well. In fact there was less visible disintegration of soil particles. From the Figure 6 (b) it is evident that there was reinforcement of the polymeric fibres. There were also voids which could cause profound effect on the compression strength. The morphology stabilized mud blocks with 0.5% of polypropylene mono filament fibres S10 is shown in figure 6(c). The reinforcement of fibre was clearly visible between the soil particles and also voids. The voids may tend to discontinue soil particles that make the soil matrices weaker and eventually lead to high damage compared to S5 and S15. In the case of S15, (figure 6 (d)) there were no voids and surface morphology looked dense. The absence of voids and dense reinforcement in the case of polystyrene reinforced mud blocks could be the reason for the better performance of the S15. Similar kind of study was carried out [23] using Sisal, a natural fibre and the results were correlated with strength, porosity and density. This study makes us to understand the changes in internal structure of mud blocks due to the compression stress, stabilizers and fibres.

5. Conclusion

The main drawback of the black cotton soil based mud blocks viz., the brittleness, lack of strength and poor damage resistance were overcome by indigenous use of polymeric fibres as a reinforcement materials and the natural binder viz., *Terminalia chebula* in the making of stabilized mud blocks. Of the three polymeric fibres used, polystyrene fibre reinforcement has yielded mud block with greater compression strength. At a concentration of 0.5% by weight of the dry soil, polystyrene could produce compression strength of 13.75 MPa. The aromatic nature of the polystyrene has resulted in the lowest water absorption of 4.1%. This could be attributed to the non-polar nature of the polymeric fibre combined with proper compaction which has resulted in the reinforcement in both horizontal and vertical directions holding the soil matrix. Further the spray erosion test indicated the attrition rate of only 1 mm in a maximum period of 2 hr. The surface morphological studies revealed the densely packed structure indicating the absence of voids and reinforcement of the polymeric fibres. Thus from the results of the study, it is certain that polymeric fibre reinforcement could increase the stability and enhance the durability of the black cotton soil based mud blocks. The results of the present study are bound to have impact on the use of stabilized mud blocks in the domain of cost effective building technology.

6. References

[1] Hanifi Binici, Orhan Aksogan, Tahir Shah, "Investigation of Fibre Reinforced Mud Brick as a Building Material" Construction and Building Materials, 19, (2005) 313–318.

[2] Hanifi Binici, Orhan Aksogan, Derya Bakbak, Hasan Kaplan, Bilge Isik "Sound Insulation of Fibre Reinforced Mud Brick Walls" Construction and Building Materials 23, (2009), 1035–1041.

[3] Peter Donkor, Esther Obonyo, Fabio Matta, and Ece Erdogmus "Effect of Polypropylene Fiber Length on the Flexural and Compressive Strength of Compressed Stabilized Earth Blocks" Construction Research Congress, (2014), 661 – 670.

[4] Robert Medjo Eko, Engelbert Dieudonné Offa, Thierry Yatchoupou Ngatcha, Lezin Seba Minsili, "Potential of Salvaged Steel Fibers for reinforcement of Unfired Earth Blocks" Construction and Building Materials, 35, (2012), 340 – 346.

[5] Chee–Ming Chan "Effect of Natural Fibres Inclusion in Clay Bricks: Physico-Mechanical Properties" International Journal of Civil, Environmental, Structural, Construction and Architectural Engineering, 5, 1(2011) 7 – 13.

[6] C K Subramania Prasad, E K Kunhanandan Nambiar, Benny Mathews Abraham "Plastic Fibre Reinforced Soil Blocks as a Sustainable Building Material" International Journal of Advancements in Research & Technology, 1, 5 (2012), 1 – 4.

[7] I.S. 3036 Laying Lime Concrete for a Waterproofed Roof Finish - Code of Practice, Bureau of Indian Standard, New Delhi. (1992), 2 – 3.

[8] K.Mahendran, N.P.Vignesh "A Study on the Influence of Soil Properties and Additives on the Strength of Mud Blocks" International Journal of Applied Engineering Research, 10, 53 (2015), 1 – 7.

[9] C.K. Subramania prasad, Benny Mathews Abraham, E.K. Kunhanandan Nambiar "Sorption Characteristics of Stabilised Soil Blocks Embedded with Waste Plastic Fibres" Construction and Building Materials, 63, (2014), 25–32.

[10] M.G.Sreekumar , Deepa G Nair "Stabilized Lateritic Blocks Reinforced With Fibrous Coir Wastes" International Journal of Sustainable Construction Engineering & Technology, 4, 2, (2013), 23 – 32.

[11] C. Galán-Marína, C. Rivera-Gómez, J. Petric "Clay-based Composite Stabilized with Natural Polymer and Fibre" Construction and Building Materials, 24, (2010), 1462 –1468.

[12] Sung-Sik Park "Unconfined Compressive Strength and Ductility of Fiber-reinforced Cemented Sand" Construction and Building Materials, 25, (2011), 1134–1138.

[13] Roohollah Bagherzadeh, Hamid Reza Pakravan, Abdol-HosseinSadeghi, Masoud Latifi, Ali Akbar Merati "An Investigation on Adding Polypropylene Fibers to Reinforce Lightweight Cement Composites (LWC)" Journal of Engineered Fibers and Fabrics, 7, 4, (2012), 13 – 21.

[14] Vinu Prakash, Amal Raj, Aravind S, Basil Mathew, Sumith V R "Studies on Stabilized Mud Block as a Construction Material" International Journal of Innovative Research in Advanced Engineering, 1, 3, (2016), 19 – 24.

[15] Peter Walker and Trevor Stace "Properties of Some Cement Stabilised Compressed Earth Blocks and Mortars" Materials and Structures/Matériaux et Constructions, 30, (1997), 545-551.

[16] Fetra Venny Riza, Ismail Abdul Rahman, Ahmad Mujahid Ahmad Zaidi "A Brief Review of Compressed Stabilized Earth Brick (CSEB)" Proceedings of International Conference on Science and Social Research (2010), 1011 – 1016.

[17] A Guettala, H. Houari, B. Mezghiche and R. Chebili "Durability of Lime Stabilized Earth Blocks" Courrier du Savoir, 2, (2002), 61-66.

[18] B. V. Venkataramareddy, K. S. Jagadish "Spray Erosion Studies on Pressed Soil Blocks" Building and Environment, 22, 2, (1987), 135-140.

[19] B. V. Venkataramareddy "Stabilised Soil Blocks for Structural Masonry in Earth Construction" Woodhead Publishing Limited, (2012), 324-363.

[20] I.S. 1725 Specification For Soil Based Blocks Used In General Building Construction, Bureau of Indian Standards. New Delhi, (1982), 3 – 6.

[21] Esther Obonyo, Joseph Exelbirt and Malarvizhi Baskaran "Durability of Compressed Earth Bricks: Assessing Erosion Resistance Using the Modified Spray Testing" Sustainability, 2, 2010, 3639 – 3649.

[22] Sanchay Patel, Ram Parvesh Prasad "Study of Behaviour of Stabilized Mud Block and Burnt Clay Brick Masonry and Stabilized Rammed Earth Walls" International Journal for Scientific Research & Development, 4, 7, (2016), 780 – 783.

[23] Saul Sitati Namango "Development of Cost-Effective Earthen Building Material for Housing Wall Construction: Investigations into the Properties of Compressed Earth Blocks Stabilized with Sisal Vegetable Fibres, Cassava Powder and Cement Compositions" A Doctoral Dissertation, Brandenburg Technical University Cottbus (2006).

Acknowledgment

Authors place on record the financial assistance received from UGC, New Delhi in the form of Major Research Project. Authors also thank the authorities of the Gandhigram Rural Institute (DU), Dindigul, India for their support and encouragement.

Study of Fibre Reinforced Concrete Using Sustainable Materials

Dr. Uma Maguesvari M [a,*], Muthaiyan P [b], Yugasini S [b], Ammaiappan M [c]

[a] *Associate Professor, Department of Civil Engineering, Rajalakshmi Engineering College, Chennai and 602 105, India.*

[b] *Assistant Professor, Department of Civil Engineering, Rajalakshmi Engineering College, Chennai and 602 105, India.*

[c] *Assistant Professor (SS), Department of Civil Engineering, Rajalakshmi Engineering College, Chennai and 602 105, India.*

umamaguesvari@gmail.com

Abstract. Our current generation is a pillar of this developing world in construction field, which also leads to decrease in natural raw materials and increase in production of many industrial waste materials creating an impact to living life through flaming and landfilling. Concrete is one of high production material in our daily life and have most probable way to substitute the waste materials. In this present work waste material are used for replacement of materials in developing the fibre reinforced concrete. Workable and mechanical characteristics of fibre reinforced concrete have assessed by substitution of coarse aggregate with recycled aggregate at(0%,10%,20%,30%, and 40%), and fine aggregate as M- sand, waste foundry sand and bottom ash and binder is replaced with silica fume 7% in addition to that fibre content as 0%,0.5%,1% and 1.5% of micro steel fibre and Basalt Fibre by weight of cement. Workability decreases with rise in incorporation of recycled aggregate and also it diminishes workability with the inclusion of fibres. The concrete mixture with replacement of coarse aggregate including micro steel and basalt fibre content have exhibited higher mechanical strength.

Key Words: Silica fume; Bottom ash; Foundry sand; Compressive strength; Sustainability

1. Introduction

A concrete has high crushing strength but less in tension, it is one of the most durable building materials. The material is very cost effective in its production, fluid in its beginning has an ability to mould in any shape and desired strength can be achieved. It has a good resistance to temperature, wind, water and cost to maintain is also low. Despite of many benefits, the concrete get failure due to its brittle in nature. The present ingredients in the concrete has an ability to resist the compressive force to some extent. While carrying the tension force the linkage of the binder and aggregate get separated. The disconnection of the cement material and aggregate cause failure to the structure as propagation of crack.

Now a days the traditional natural concrete materials are getting reduced and industrial waste products are increasing more. By disposing the waste materials in land also it creates impact and shortage of land. The incorporation of fibre in matrix has a greater influencing property of concrete and various research proved that fibres have increased the hardened characteristics of concrete such as crushing strength, split tensile strength, impact resistance, flexural strength [1]. The inclusion of waste material and fibre has an effect in crushing strength of concrete. It has shown that increase in compressive strength at lower fibre content, whereas decrease in compressive strength at higher content of fibres addition [2]. Fibre type and aspect ratio of fibre also influences the mechanical characteristics on concrete. In which Basalt Fibre is melted and extruded from basalt rock, where else Glass fibre are formed in a process in which molten glass is drawn in the form of filaments. By comparing both the fibres Basalt fibres has higher tensile strength then Glass fibre [1,3,4]. Investigation have been carried on waste materials like Bottom Ash, it represents that at lower content of 15-20% Bottom ash give more compressive strength, Usage of fibre reduces the workability of concrete [5-8]. It has attempted to utilize the bottom ash as a substitution of fine aggregate in making cellular concrete [9]. Studies have carried on Waste foundry Sand and Coal Cinder CC, it is proven that workability is been decreasing by increase of CC and WFS because water absorbed by fine particles, and crushing strength of WFS is increased by up to 30% of substitution by fine aggregate after that strength reduces gradually [10-13]. Silica fume one of binding material which has a pozzolanic property that replaced for cement at an optimum of 7-12% [14]. Field studies have been carried out for usage of recycled aggregate on foundation system, municipal side walk construction and performance studies have been carried out and collate with normal concrete [15]. Mechanical and transport properties have evaluated the influence of incorporation of recycled aggregate considering various curing conditions [16]. Hardened and durability characteristics have investigated on the recycled aggregate with the use of ternary binder and optimum mixes were arrived with the crushing strength of 52 MPa [17]. Based on the above perspective substitution of coarse aggregate with recycled aggregate at 0%,10%,20%,30%, and 40% and fine aggregate as M- sand, waste foundry sand and bottom ash and binder are replaced with silica fume 7% by weight of cement were used to determine the workable and harden characteristics of concrete. It leads to sustainability, limited negative impact on natural resources and avoids depletion or destruction of natural resources.

2. Materials and mix proportion

2.1. *Cement*
OPC of grade 53 confirming to IS 12269-1987, used in this present work, which has a specific gravity of 3.15 and 30% of consistency.

2.2. *Silica fume*
It is an ultrafine powder with a specific gravity of 2.63. The content of silica fume used as 7% for the weight of cement.

2.3. *M Sand*
Manufactured sand produced from crushing of granite has a specific gravity of 2.74. The sand used has grading conforming to zone II as per IS 383:1970.

2.4. *Bottom Ash*
Bottom Ash (BA) is by-product of coal which is produced in thermal power plant as a residue in boiler. This product is been collected from Neyveli Lignite Corporation Tamil Nadu. It was observed that specific gravity is 2.4 and 1396 kg/m^3 per unit weight.

2.5. *Waste Foundry Sand*

Waste Foundry Sand (WFS) is waste product from metal casting industry which is obtained from Coimbatore, which has a specific gravity of 2.6 and 1910 kg/m³ of per unit weight.

2.6. *Gravel*
Blue metal aggregate is used as a coarse aggregate in this work as per IS 2386 (part 3). Which is obtained from locally available quarry of maximum size of 20mm is used for this investigation and which has a specific gravity of 2.74.

2.7. *Recycled aggregate*
Recycled concrete aggregates produced from typical concrete. Recycled concrete aggregates contain crushed aggregate with hydrated cement paste. The specific gravity of Recycled aggregate as 2.46 with the maximum size of 20mm.

2.8. *Water*
Pure water is used for making concrete and curing as per IS 456:2000.

2.9. *Fibres*
Micro Steel fibre of 12mm length and 0.2mm diameter and Basalt fibre of 12 mm length and 13-20μm of diameter, were used in investigation.

2.10. *Mix Proportion*
This investigation has been carried into two phases. In the first phase binder content as 400 Kg/m^3 with the replacement of 7% of silica fume, fine aggregate as the combination of M- sand, Foundry sand and Bottom ash with the percentage of 60:20:20, Coarse aggregate with the replacement of recycled aggregate from Concrete demolition waste with percentage from 0 to 40% with the increment of 10 percentage. The ratio is been arrived for M 40 grade of concrete from the mix design as per IS 10262: 2019 is 1:1.8:2.946 at 0.45 w/c. 5 mixes were used to find the workability characteristics from slump test, mechanical characteristics such as crushing strength and density. One optimum mix is identified from the above mixes. In phase two, from the identified mix two types fibers such as steel fibre and basalt fibres with the percentage of 0.5, 1 and 1.5% are introduced in concrete, 6 mixes are used to measure the workability, crushing strength and split tensile strength of concrete. Totally 5 mixes without fibre and 6 mixes with fibers.

3. Result and Discussion

3.1. *Fresh and Hardened characteristics without fibre*

3.1.1. Workability. The homogeneity of concrete is tested by slump cone. It is very useful in site construction for making the concrete in good flow by adjusting the water cement ratio. The slump value for variation of replacement of recycled coarse aggregate are shown in Fig. 1. It shows that there is a falling pattern of slump value with the rise in recycled aggregate substitution. There is a gradual reduction in slump value up to 20% replacement then there is a sudden dip in slump value after 20%.

3.1.2. Density. The variation of density for various substitution of coarse aggregate are shown in Fig.2. density of recycled concrete varies from 2360 to 2435 Kg/m3 for aggregate replacement from 0 to 40% with the increment of 10% increase in recycled concrete aggregate. It represents that there is a gradually decreasing trend with the increment of recycled concrete aggregate. This may be due less specific gravity of recycled aggregate and also increased porosity compared to similar virgin aggregates

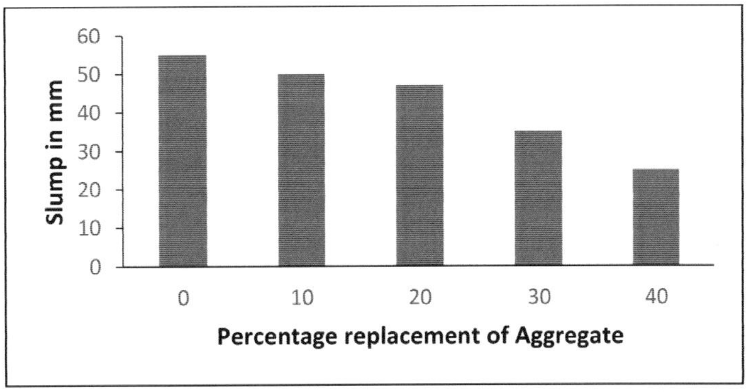

Fig 1: Variation of slump Value for Percentage of replacement of Coarse Aggregate

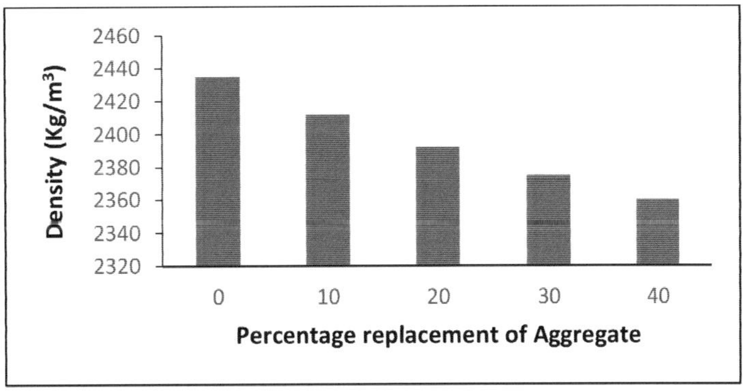

Fig 2: Variation of Density for Percentage of replacement of Coarse Aggregate

3.1.3. Compressive Strength. The crushing strength is found from the average of 3 specimens at 7 and 28 days of cured concrete. The size of the cube 150mm and test is carried out as per IS 516-1959. The variation of crushing strength at the age of 7 and 28 days for various substitution of coarse aggregate as recycled aggregate are given in Fig 3. It is noticed that substitution level of recycled aggregate rise in concrete reduces the crushing strength of concrete. There is a marginal decrease in percentage up to 20% replacement of recycled aggregate with increase in replacement 30% and 40% which is more than 10% reduction in compressive strength.

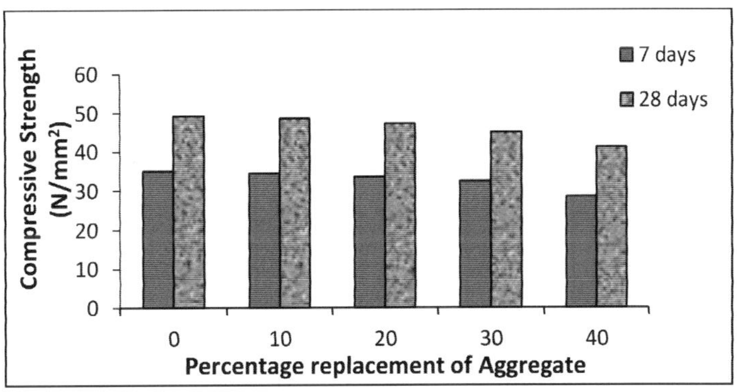

Fig 3: Variation of 7- and 28-days compressive strength for Percentage of replacement of Coarse Aggregate

3.2. Fresh and Hardened characteristics with fibre

3.2.1. Workability of fibre reinforced Concrete. The observed slump Range is shown in Fig.4 with the incorporation of steel and basalt fibre at 0.5%, 1%, 1.5%. The influence of fibres in concrete also reduces workability to some extent among the fibres because it will not allow concrete to segregate, flow, and also high content of fibres which leads to wrap the cement paste around it. Pattern of slump value is same for both the fibres. Slump value of steel fibre is less when compared to basalt fibre of various percentage of fibre addition. The slump value varies between 10 to 35 mm irrespective of the type of the fibre added.

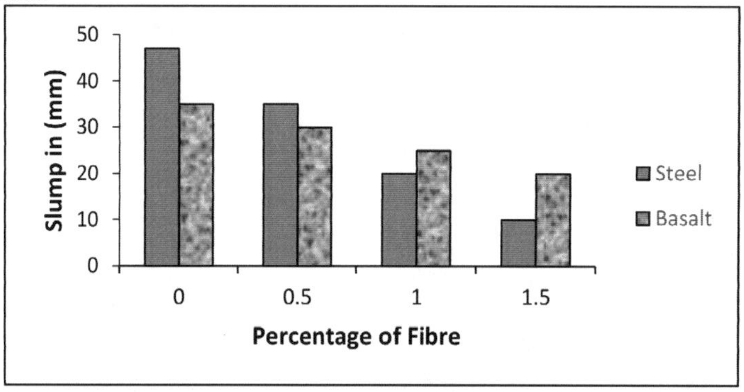

Fig 4: Variation of slump Value for Percentage of steel and Basalt fibers.

3.2.2. Compressive strength of fibre reinforced Concrete. The change of crushing strength for 7 and 28 days of different fibre percentage of steel and basalt fibres are shown in Fig 5. The crushing strength differ from 47.32 to 50.12 MPa irrespective of fibre type (steel and basalt fibre). There is a gradually increasing compressive strength with the incorporation of steel and basalt fibre in concrete. Percentage rise in crushing strength upto 1% of steel fibre was 5.5% and 4.6% for basalt fibre. With further increment in fibre percentage for type of fibre attained of improvement in compressive strength is less than 1% only. By comparing the compressive strength with steel fibre and basalt fibre, steel fibres exhibited more compressive strength than basalt fibre.

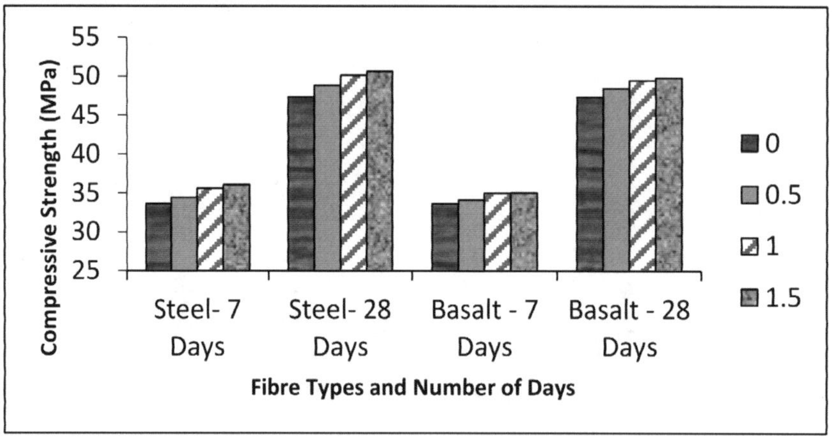

Fig 5: Variation of Crushing strength for Percentage of steel and Basalt fibers.

3.2.3. Split Tensile strength of fibre reinforced Concrete. The variation of split tensile strength with the different percentage of fibres are given in Fig 6. Split tensile strength varies from 4.94 to 7.25 MPa for steel fiber and 4.94 to 6.94 MPa for Basalt fibres for fibres from 0 to 1.5% with the increment 0f 0.5% of fibre content. There is an increasing patten of attainment of split tensile strength. Percentage attainment of increment of strength is maximum up to 1% with further increase in percentage of fibre of about 0.5%, percentage attainment is comparatively less than attainment up to 1%. Above behaviour is similar for both steel fibre and basalt fibre. By comparing the attainment of split tensile strength steel fibers shows better than basalt fibre.

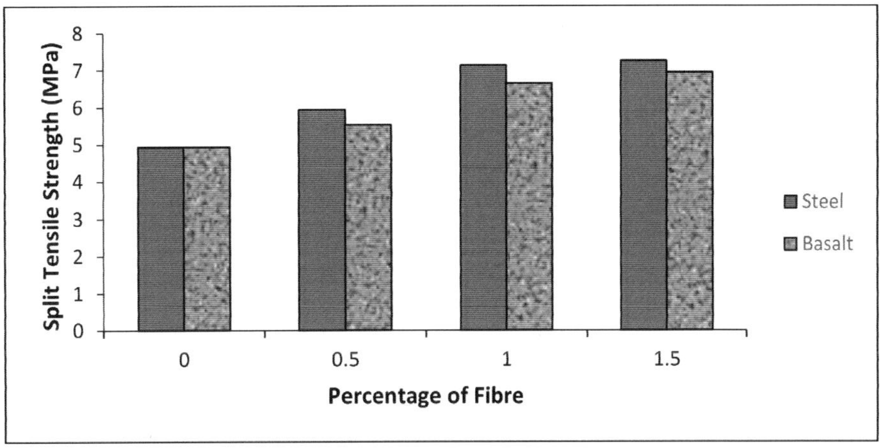

Fig 6: Variation of Split Tensile strength for Percentage of steel and Basalt fibers.

3.3. Identification of mix and sustainability

Based on the observation from the crushing strength and split tensile strength mix with 1% steel fibres and basalt fibre have identified for further progress of the research and test. In order to preserve the natural resources, needless losing of natural assets ought to be limited and regulated. Formulation and implementation of appropriate waste management plan can decrease C and D waste. Recycled material can be reuse and greater natural assets is conserved for our subsequent generations.

3.4. Ultra-sonic Pulse Velocity

Ultrasonic pulse velocity test is used to measure the integrity of concrete as per IS 13311 (Part 1) – 1992. After 28 days of curing before conducting the test for mechanical properties UPV values was found on all cubes of 100mm which is as shown in Fig. 7. The quality of concrete is examined using UPV and results have exhibited excellent quality of concrete for observed cubes specimen shown in Table 1.

Fig 7: Ultrasonic Pulse Velocity for Cube specimen

Table 1 Ultra-Sonic Pulse Velocity for 1% fibres content

Specimen	Velocity (m/sec)	Quality of concrete
Control mix	4870	Excellent
Steel fibre 20%	5010	Excellent
Basalt fibre 20%	4940	Excellent

4. Conclusion

The following conclusion were arrived from the fibre reinforced concrete using sustainable materials.

- Slump value of concrete reduces with the rise in percentage substitution of recycled concrete aggregate.
- Addition of Steel and Basalt Glass fibres decreased in workability of concrete gradually with the increment of fibre percentage.
- Density and crushing strength of concrete diminish with the rise in substitution of recycled coarse aggregate.
- Mix with 20% replacement of recycled concrete aggregate has identified for further work with fibres.
- Inclusion of steel and basalt fibres increases the crushing strength and split tensile of concrete.
- Steel fibres shows better strength than basalt fiber concrete with same percentage of fibre 1%.
- Use of recycled aggregate as coarse aggregate and fine aggregate as M- sand, foundry sand and bottom ash reduce the depilation of natural resources in terms of sustainability.

References

[1] Ahmet B Kizilkanat, Nihat Kabay, Veysel Akyuncu, Swaptik Chowdhury and Abdullah H. Akca 2015 Mechanical properties and fracture behavior of basalt and glass fiber reinforced concrete: An experimental study Construction and Building Materials 100 pp 218-224

[2] Piotr Smarzewski and Danuta Barnat-Hunek 2016 Mechanical and durability related properties of high performance concrete made with coal cinder and waste foundry sand Construction and Building Materials 121 pp 9-1

[3] Mehmet Emin and Arslan 2016 Effects of basalt and glass chopped fibres addition on fracture energy and mechanical properties of ordinary concrete: CMOD measurement Construction and Building Materials 114 pp 383-389

[4] Suchita Hirde and Sagar Shelar 2017 Effect of Basalt Fiber on Strength of Cement Concrete International Journal of Current Engineering and Technology Vol 7 No 2 pp 600-602

[5] Navdeep Singh, Mithulraj M and Shubham Arya 2018 Influence of coal bottom ash as fine aggregates replacement on various properties of concretes: A review Resources, Conservation & Recycling 138 pp 257-271

[6] P Aggarwal YAggarwal and S M Gupta 2014 Effect of Bottom Ash as Replacement of Fine Aggregates in Concrete Asian Journal of Civil Engineering Building And Housing Vol 8 pp 49-62

[7] Dinesh Kumar G et al 2016 An Experimental Study on Partial Replacement of Fine Aggregate with Coal Bottom Ash in Concrete IJSAE Vol. 2 pp 39-49

[8] Akkadath Abdulmatin Weerachart Tangchirapat Chai Jaturapitakkul 2018 An investigation of bottom ash as a pozzolanic material Construction and Building Materials 186 pp155-162

[9] Patchara Onprom, Krit Chaimoon and Raungrut Cheerarot 2015 Influence of Bottom Ash Replacements as Fine Aggregate on the Property of Cellular Concrete with Various Foam Contents Advances in Materials Science and Engineering Vol 2015 381704

[10] Piotr Smarzewski and Danuta Barnat-Hunek 2016 Mechanical and durability related properties of high performance concrete made with coal cinder and waste foundry sand Construction and Building Materials 121 pp 9-1
[11] Rafat Siddique and Gurpreet Singh 2011 Utilization of waste foundry sand (WFS) in concrete manufacturing Resources Conservation and Recycling 55 pp 885-892
[12] Vema Reddy Chevuri and S Sridhar 2015 Usage of Waste Foundry Sand in Concrete SSRG-IJCE vol 2 pp 5-10
[13] Amrutha K et al 2017 Partial Replacement of Fine Aggregate by Used Foundry Sand in Concrete IRJET Vol 4 165-169
[14] Chetan S Kankariya Hitesh S Patil Sudarshan Bhamre Mayur D Mhasde 2017 Review Article: Study And Optimization Of Silica Fume in Concrete International Journal of Recent Trends in Engineering & Research Vol 03 Issue 05 pp 562- 565
[15] Humera Ahmed Mohammed Tiznobaik Sumaiya B Huda M Shahidul Islam M Shahria Alam 2020 Recycled aggregate concrete from large-scale production to sustainable field application Construction and Building Materials 262 119979
[16] Ahmed Shaban Abdel-Hay 2017 Properties of recycled concrete aggregate under different curing conditions HBRC Journal 13 271–276
[17] O E Babalol P O Awoyera M.T. Tran D H Lea O B Olalusi A Viloria D. Ovallos-Gazabon 2020 Mechanical and durability properties of recycled aggregate concrete with ternary binder system and optimized mix proportion J Mater Res Tech 9 3 pp 6521–6532

IOP Conf. Series: Materials Science and Engineering 1026 (2021) 012012 doi:10.1088/1757-899X/1026/1/012012

EXPERIMENTAL STUDY ON PARTIALLY REPLACED CONCRETE WITH AUTOMOBILE INDUSTRIES WASTES

Ms.M.Kalpana[1] and Mr.G.Sudharson[2]

[1]Associate Professor, Department of Civil Engineering, Saveetha Engineering College, Chennai

[2]Assistant Professor, Department of Civil Engineering, Saveetha Engineering College, Chennai

Abstract. In recent days, the major threat to our environment is waste management. As a contribution to save our environment many research works are happening to utilize waste in their manufacturing process to reduce waste as well as reducing the cost involved in the process. Waste utilization will partially help us to protect our environment. In this paper, for manufacturing concrete with partially replaced fine aggregate is experimented. The objective is to reuse the locally available polypropylene fibre generated as waste by automobile industries. The adhesion between polypropylene fibre and cement matrix is considered as an important factor. To give proper matrix between the fibre and cement, shredded polypropylene fibre is used as partially replaced fine aggregate. The replacement percentages used are 20%, 40% and 60%. The specimens were prepared with water cement ratio 0.5 and tested after 7days and 24 days of curing period. The influences of polypropylene fibre in concrete in fresh and hardened state are compared with conventional concrete properties. The results obtained shows increase in the strength and workability property for 20% replacement of polypropylene fibre as fine aggregate in concrete. Hence the disposal of polypropylene waste from automobile industry can effectively be used in shredded form as partial replacement of fine aggregate in concrete.

Keywords: Polypropylene fibre, Automobile waste, Fresh and Hardened state

1. INTRODUCTION

A large amount of waste as plastic, nylon, rubber are produced by the manufacturing industries like automobile parts, household goods, construction projects, industrial wastes etc. In that the automobile waste plastics are non-biodegradable and needs to be recycled. Recent studies have shown that several waste materials can be profitably used to manufacture low cost structural and non-structural components in construction industry. Concrete have a high compressive strength but low tensile strength. To overcome the disadvantages like strain softening the waste fibre can be used. The performance of cement composite depends on the fibre properties, surface texture of fibre, type of fibre, cross section of fibre and the matrix and fibre/ matrix interface properties. The adhesion property between the fibre and the cement matrix is an important aspect governs the performance of cement composite. Polypropylene fibre have been used in cement composite to control cracking due to shrinkage, to improve the toughness of the materials, to improve impact resistance, and to increase the energy absorption capacity of the material. The polypropylene fibre improves mix cohesion, pump ability over long distance, freeze-thaw resistance, impact resistance, increase resistance to plastic shrinkage during curing. The polypropylene fibre obtained from automobile waste is shredded into irregular shapes by means of shredder. The manufacturing process of polypropylene fibre from automobile waste to shredded form is shown in fig 1.1.

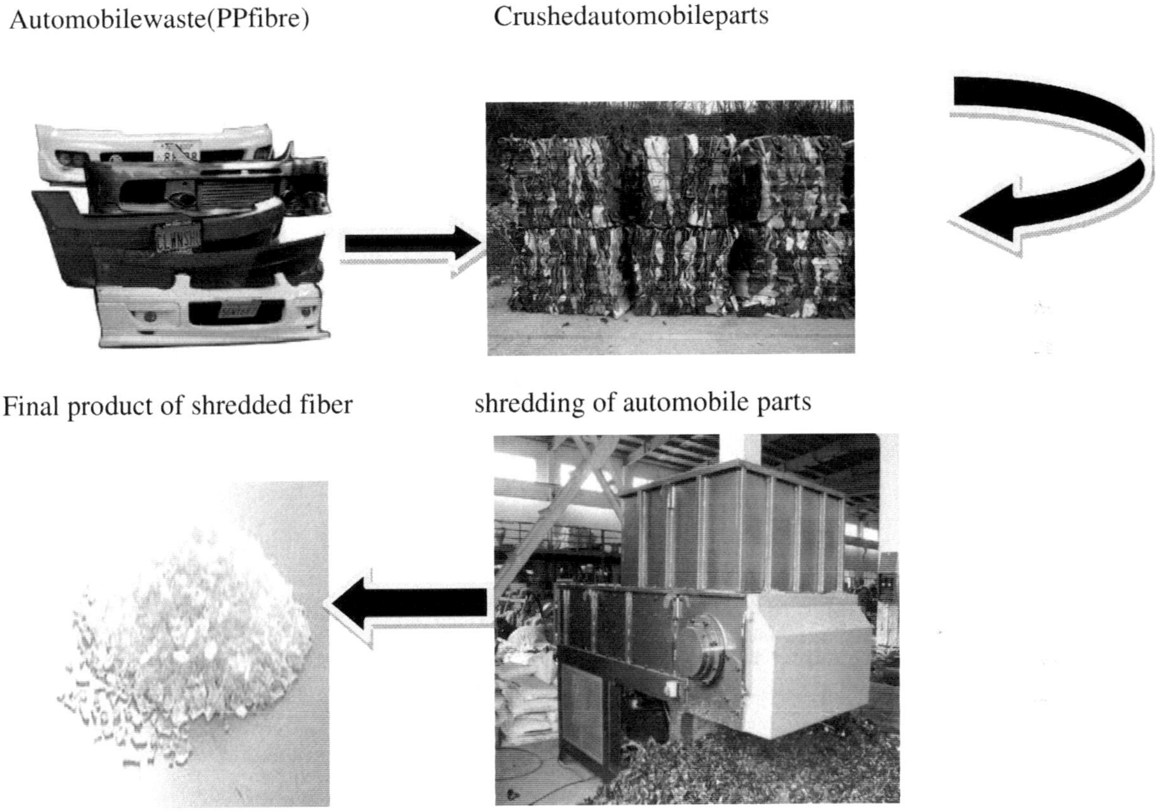

Fig 1.1 Shredded Polypropylene waste from source

2. LITERATURE SURVEY

Ammar [1] et al, suggested that the mechanical characteristic of no-fine concrete with or without polypropylene fibre are investigated. The results show that the addition of polypropylene fibre in no-fine concrete can improve the mechanical properties, especially the tensile and compressive strength by 19% and 17% respectively compared with the concrete without fibres, and a slight effect on the density. The author concluded that polypropylene fibre actually help to inhibit the formation of crack due to both plastic shrinkage and dry shrinkage. The polypropylene fibre help to maximize the intrinsic strength of concrete and also improve mix cohesion to reduce the bleeding of water.

T.Abadjieva [2] et al, no fine concrete is the form of lightweight porous concrete, the advantage of this type of concrete are lower density, low cost due to less cement content, low thermal conductivity, low drying shrinkage, no segregation and capillary movement of water. The concrete mixes made without fine aggregate and single sized coarse aggregates fraction 13.5 -19mm. The author concluded that the strength of no fine concrete is lower than the normal weight concrete, but sufficient enough for structural use. The suggested mixture could be used for cast in-situ walls in low rise, low cost housing, drainage layers and paving after more extensive research.

Anne Laning [3], synthetic fibres are most commonly added to concrete. It reduces the cracks formed due to early plastic shrinkage and increases the resistance to impact, abrasion and toughness. The fibres can also be added to precast concrete to improve resistance. It is added to pumped concrete to improve cohesiveness. Fibres are added to shotcrete to reduce rebound and material waste. The author concluded that replacing the wire mesh as secondary reinforcement it inhibits the early shrinkage cracking.

Faisal Fouad Wafa [4], said that the concrete is weak in tension and is brittle in nature. Continuous reinforcement used in concrete helps to increases the strength and ductility only when done with proper workmanship. When crack occurs due to shrinkage, the randomly oriented fibres will arrest the formation of cracks. Fibres also arrest the propagation of cracks. It improves the strength and ductility property of concrete. The expected failure modes of fibre are bond failure between fibre and matrix or material failure. The author concluded that workability problem was not encountered when hooked fibre are used up to 1.5 percent in the concrete mix. The straight fibre produces balling at high fibre content and requires special handling procedure to improve the ductility of concrete. Fibre reinforced concrete controls cracking and deformation under impact load. The results are much greater than plain concrete and increased the impact strength of 25 times.

M.A.Islam [5] et al, has discussed about the effectiveness of using polymer modified concrete (PMC) as a construction material. A detailed literature survey reveals that the concrete property can be varied by varying the nature and concentration of polymer materials. They mentioned about the applications such as protecting the constructed structures from floods, earthquakes, saline water and landslides at embankments. The author concluded that PMC possess remarkable potential due to its features, properties and applications. The material can effectively respond to the needs of current and future construction works. Infrastructure can be repaired by using polymer modified concrete. It will be extremely feasible and with low maintenance requirements.

Kolli. Ramujee [6], the compressive strength and split tensile strength of samples were compared for various proportions of fibre like 0.5%, 1.0%, 1.5% and 2.0%. There was a reduction in slump value was noted by the author with increasing proportion of fibre content. The author concluded that with 1.5% of fibre content produces optimum result when compared with other sample. The optimum increase in compressive strength was noted as 34% and split tensile strength as 40% when compared with conventional concrete.

Mr. Mehul J. Patel [7] et al, said that the effect of various proportion polypropylene fibre on the properties of high strength concrete was carried out to analyze its effects on compression, tensile, flexural, shear strength and plastic shrinkage cracking. To find the optimum percentage of polypropylene fibre content in the concrete mix, test was carried out for 0.5%, 1%, 1.5% of fibre at different age level of concrete. The author concluded that the polypropylene fibre can be used as partial replacement for cement along with admixtures to increase the strength of concrete. Result showed that the workability of polypropylene fibre replaced concrete is decreasing with increase in polypropylene fibre content.

H.R.Pakravan [8] et al, the fibres plays an important role in improving the cementitious composites fracture toughness. The author discussed about the adhesion between the polymeric fibre and cement matrix. The pull-out behaviour of all tested fibres was almost same for 7 and 14 days curing specimens. But the author found increase in pull-out load of all 28 days curing specimens. Results were proved that N66 fibres can undergo pull out load.

Saeed Ahmed [9] et al, the author said that the polypropylene fibre is synthetic hydrocarbon polymer so the effect of addition of various proportion of polypropylene fibre in concrete mix gives the notable increase in flexural, tensile and shear strength. The author concluded that the shrinkage cracking is reduced by 83-85% by addition of fibre in the range of 0.35 to 0.50% and the shear capacity is increased when the fibres are added but the effect of addition of polypropylene fibre does not increase the value of compressive strength was found.

M.Sivaraja [10] et al, said that the mechanical strength properties of synthetic fibres with volume fractions (0.5%, 1% and 1.5%) and aspects ratios (30, 60 and 90) were evaluated. Concrete mixed with rural waste fibres improved mechanical strength. Steel, nylon, plastic and coir fibres in mechanical strength enhancement were found more than that of other fibre. The author concluded the optimum volume fraction of fibres was 0.5 -1%. Fibre reinforcing index was most influential parameter and increase in strength of fibrous concrete is proportional to fibre reinforcing index of all proposed models exhibited correct correlation with experimental results.

D. Van Gemert [11] et al, discussed about the materials behaviour, especially in the field of admixtures, and about the curing processes. The author concluded that the polymer has no longer special construction materials that replace classical minerals or organic building materials. As the result polymer is now the vital component in the production of composite and sustainable building materials.

Youjiang Wang [12] et al, discussed about the experimental study conducted. Examined the tensile properties of reinforced concrete with acrylic, nylon and aramid fibres, in the form of random distribution by means of three different tests are compact tension, flexural and split tensile test. The authors were found that the strength was improved by the inclusion of fibres for reinforcement. The author concluded that the small volume fraction synthetic fibre (2 to 6.5%) added during preparation of a cement matrix have shown significant resistance to failure.

3. EXPERIMENTAL DATA

Polypropylene Fibre: It is hydrophobic. They do not absorb. It has no effect on concrete mixing water requirement. They are available either as fibrillated bundles or monofilaments. To produce fibrillated fibres, manufacturers extrude the polypropylene in sheets that can be stretched and slitted. The result is a mesh of interconnected fibre stands and is rectangular in cross section. With fibrillated fibres cement paste penetrates into the network of fibre filaments resulting in better mechanical anchoring to the concrete. The shredded fibres are partially replaced to fine aggregates with volume fraction of 20%, 40% and 60%.

Fibre	Specific Gravity	Length in inches	Tensile Strength in psi	Young's Modulus in ksi
Polypropylene	0.91	¼-2½	80-110	500-700
Polyethylene	0.91	½-¾	40-100	500-700
Polyester	1.34	¾-2	80-170	1450-2500
Nylon	1.16	¾-2	130	750

Table 3.1 Properties of synthetic fibre

Cement: It can be defined as the binding material possesses cohesive and adhesive properties. It makes the cement effective to unite different types of construction materials and form the compacted assembly. Ordinary Portland cement is one of the most widely used types of cement. A mixture composed of properly homogenized limestone and clay. It consists of inorganic oxides; the main one is lime (CaO) with a basic function and silica (SiO2) which is acidic in nature. The cement has the specific gravity of 3.02 and the physical properties are conforming to IS: 12269 – 1987. The cement used was Ordinary Portland cement (53Grade) with a specific gravity of 3.15. Initial and final setting times of the cement were 65 min and 197 min, respectively.

Aggregates: IS 383 & 2386 is the publication deals with specifications for Coarse and Fine aggregates from natural sources for Concrete. These specifications do not specify any limit for fineness modulus to be used in concrete. It divides the sand in four zones i.e. from Zone I to Zone IV. Zone I–Sand being very coarse and Zone 4 sand is very fine. It is generally recommended by code to use sands of zones I to Zone III for Structural concrete works. The fine aggregate shall have not more than 45% passing any sieve and retained on the next consecutive sieve and its fineness modulus will not be less than 2.3 and not more than 3.1.

The River sand was used as fine Aggregate and conforming to Zone III with specific gravity of 2.6, Bulk density of 1680 kg/m3. Coarse Aggregates are Crushed angular granite metal from a local source was used. The Specific gravity was 2.70, Bulk density of 1450 kg/m3 with void percentage 1.5. The physical properties are confirming to IS: 2368-1968.

Water: A concrete mix contains the minimum amount of portable water that has been required for complete hydration of cement. If the concrete is fully compacted it helps to develop the maximum attainable strength at preferable age.

Mix Proportion: The Concrete mix design has been calculated for various proportions and arrived at final mix proportion. It is used for combining the initial materials. After mixing the initial materials, fibres were added. In this the concrete samples were prepared with fibre ratios of 20, 40 and 60% by volume. The ratio of water added to the cement was w/c = 0.50.

Ingredient	Quantity in Kg/m3
Water	0.745
Cement	1.49
Fine aggregate	2.02
Coarse aggregate	3.74
Water cement ratio	0.5

Table 3.2 Mix proportion of M20 grade concrete

Test to be conducted: The properties of conventional concrete and partially replaced fibre concrete are to be determined in fresh and harden concrete. The fresh concrete test's considered is slump cone test, compaction factor test, flow table test and Vee Bee consistometer test. To determine the hardened concrete properties, cube with 150x150x150mm, beam with 500x100x100mm and cylinders with 150x300mm height are to be casted. Harden concrete test to be performed are compression test, flexure test and split tensile tests cube, beam and cylinder respectively.

4. RESULTS AND DISCUSSION

COMPARISION OF SLUMP VALUES: The workability of concrete mix is mainly determined to find the type of construction, placing condition and the means of compaction required. The properties of fresh concrete, amount and condition of reinforcement, the shape and size of mould are important factor, which control workability. Workability was measured in terms of slump. The slump test results for conventional concrete and with different percentage of polypropylene fibres are given in table 4.1.

Sample	Slump Height in cm	Workability
Conventional concrete	25	Good
20% fibre replaced concrete	26	Very Good
40% fibre replaced concrete	28	Very Good
60% fibre replaced concrete	28	Very Good

Table 4.1Comparison of slump cone results

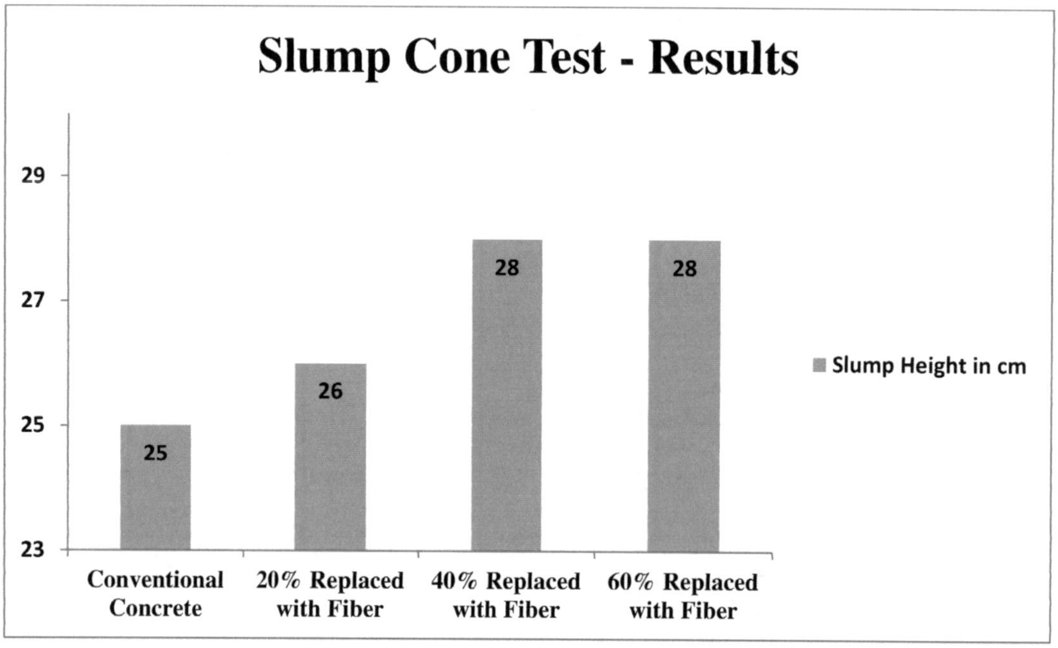

Fig 4.1Comparison of slump cone results

From the result it is observed that the workability of conventional and polypropylene fibre replaced concrete are good.

COMPARISION OF COMPACTION FACTOR VALUES: Workability of a concrete indicates the full compaction of concrete using a required amount of work. It helps to achieve the desired possible density of fresh concrete with better strength and durable structure. It helps to maintain durability throughout the job. Workability was measured in terms of compaction. The compaction factor test results for conventional concrete and different percentage of polypropylene fibres are given in table 4.2.

Sample	Density of partially compacted concrete	Density of fully compacted concrete	Compaction Factor

Conventional concrete	8.44	10.41	0.81
20% fibre replaced concrete	8.09	10.5	0.84
40% fibre replaced concrete	8.7	11.2	0.77
60% fibre replaced concrete	8.9	13.2	0.67

Table 4.2 Comparison of compaction factor values

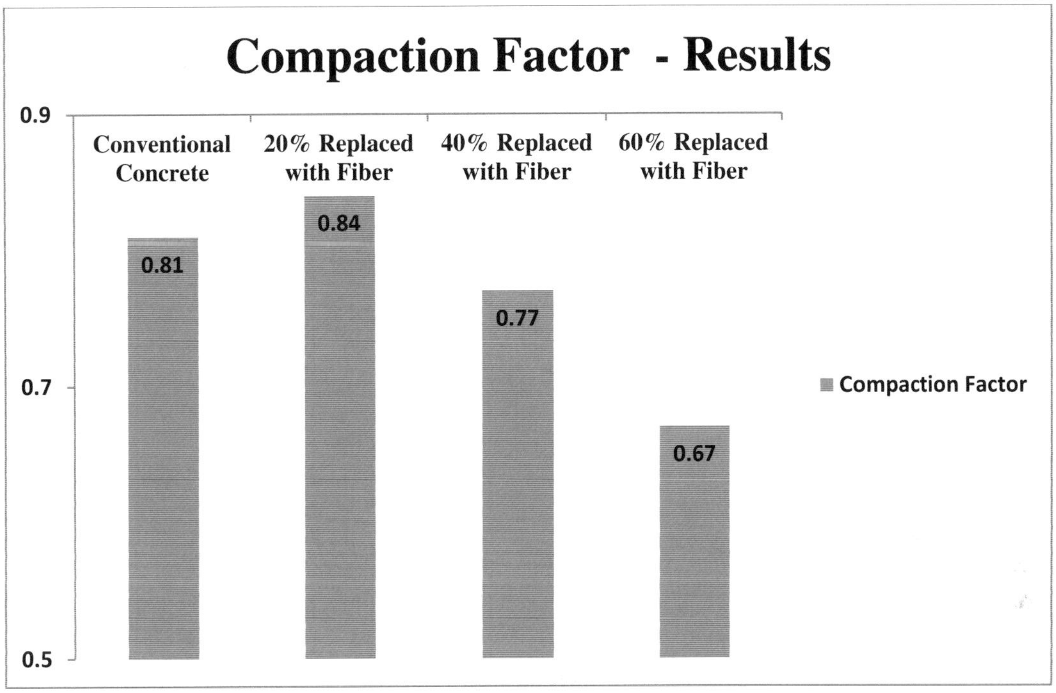

Fig 4.2 Comparison of compaction factor values

From the graph it is observed that 20% replaced concrete has a compaction factor of 84% hence it can be used for reinforced concrete structures. The compaction factor for 40% and 60% replaced fibre are 77%, 67% respectively and these can be used for road pavements.

COMPARISION OF VEE-BEE CONSISTOMETER VALUES: In general workability reduces with higher dosage of fibres compared to initial dosage used. Due to increase in addition of fibres, there is increase in amount of entrapped air voids. This is due to presence of fibres and hence it contributes to increases in air content which reduces the workability of concrete. When fibre content replacement percentage is increased, it is noted that difficulty in compaction of the mixes. It may also tend to cause improper finishing problem. The vee-bee consistometer test results for conventional concrete and different percentage of polypropylene fibres as shown in table 4.3.

Sample	Initial Reading in cm	Final Reading in cm	Slump Height in cm	Time Taken in sec
Conventional concrete	6	16	10	10
20% fibre replaced concrete	7	16.5	9.5	9.5
40% fibre replaced concrete	7.5	16.9	9.4	9.4
60% fibre replaced concrete	8	14.6	6.6	6.6

Table 4.3 Comparison of vee-bee consistometer values

According to the results both 20% and 40% replaced concrete mix gives good workability than conventional concrete. The slump loss occurs when the fine aggregates are replaced by 60% of polypropylene fibre.

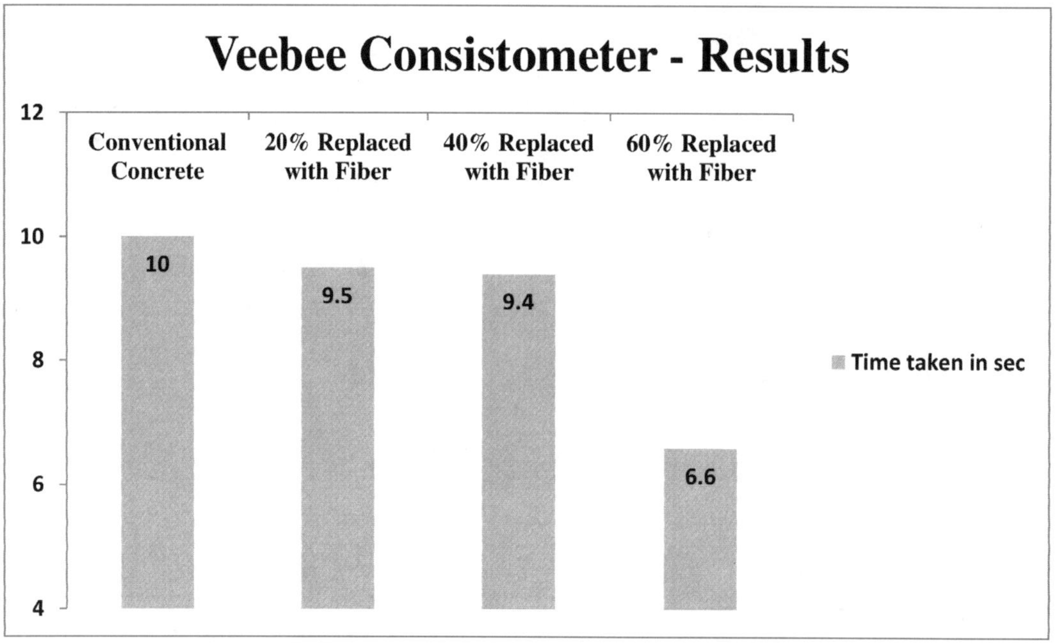

Fig 4.3 Comparison of vee-bee consistometer values

COMPARISION OF FLOW TABLE VALUES: It is used to determine consistency of fresh concrete. The concrete will not possess the required strength if the consistency of concrete is not at the desired level. If concrete is too pasty, cavities may form with in the concrete itself. Reinforcement bars may become corroded and the concrete will begin to form cracks. Cavities and cracks reduce the strength of concrete. The flow table test results for conventional concrete and different percentage of polypropylene fibres are shown in table 4.4.

Sample	Original Diameter in mm	Spread Diameter in mm	Flow %
Conventional concrete	250	470	88
20% fibre replaced concrete	250	462	85
40% fibre replaced concrete	250	460	84
60% fibre replaced concrete	250	450	80

Table 4.4 Comparison of flow table values

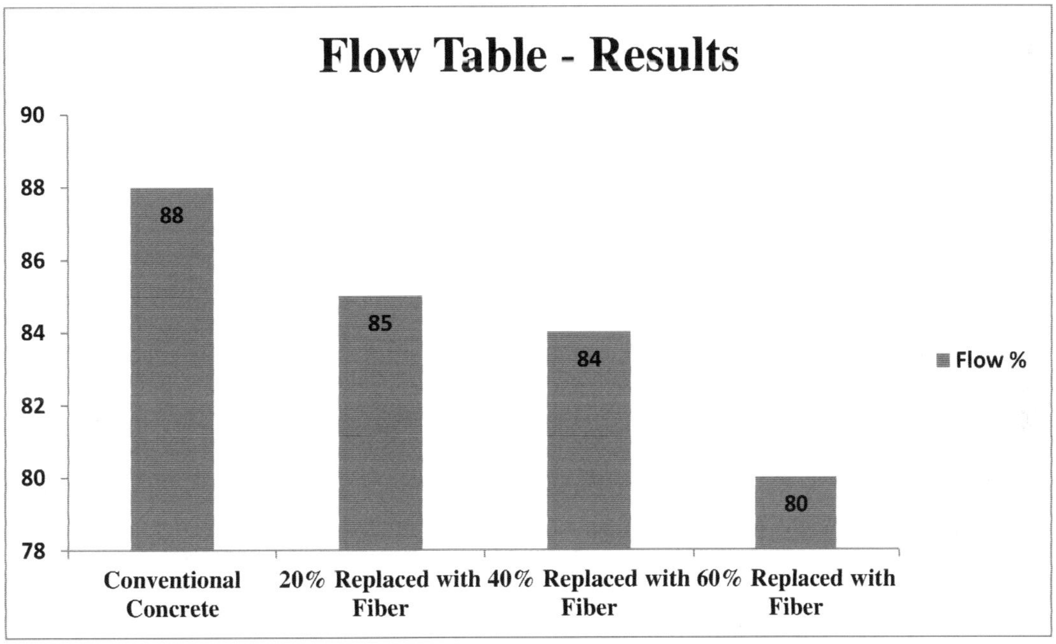

Fig4.4 Comparison of flow table values

COMPARISION OF COMPRESSIVE STRENGTH: In general, improvement in strength factor is observed due to fibre addition. It is mainly due to confinement provided by the fibre when added to concrete. The test results are shown in table 4.5.

Sample	Load in KN		Fibre Dosage in kg/m^3	Compressive Strength in N/mm^2	
	7 Days	28 Days		7 Days	28 Days
Conventional concrete	350	460	0.0	15.5	20.4

20% fibre replaced concrete	400	480	0.24	17.77	21.33
40% fibre replaced concrete	360	410	0.68	16.0	18.22
60% fibre replaced concrete	320	340	1.14	14.3	15.1

Table 4.5 Comparison of compressive strength

It has been observed that the addition of 20% of polypropylene fibres to the mix increases both 7 days and 28 days compressive strength than conventional concrete. Also it is observed that 60% fibre replaced concrete shows reduction in compression strength.

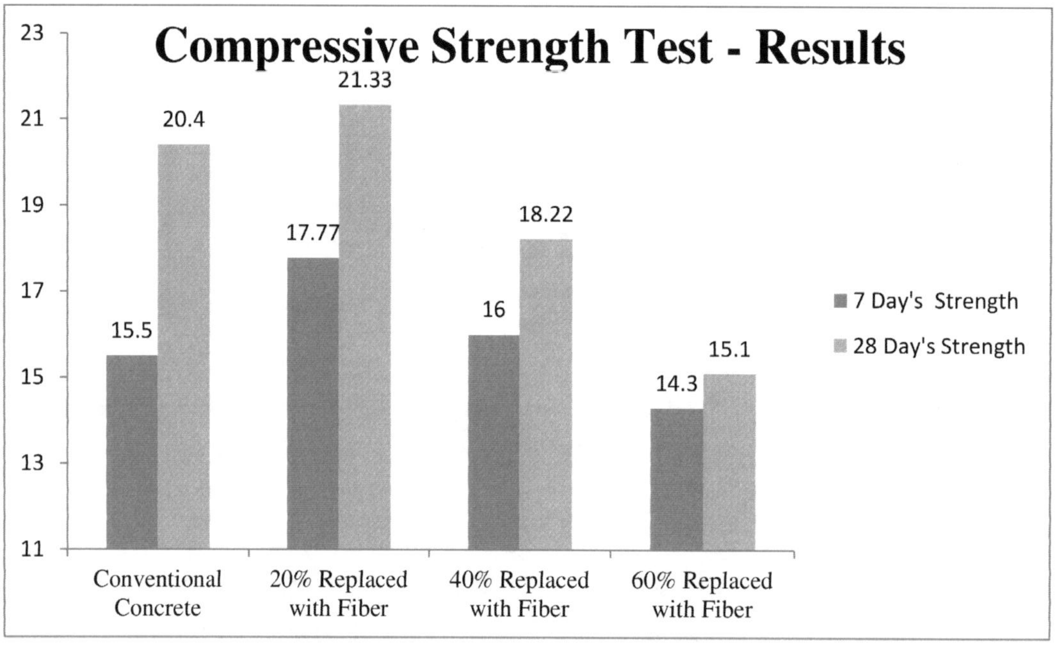

Fig 4.5 Comparison of compressive strength

COMPARISION OF SPLITT ENSILE STRENGTH: In general, the split tensile strength test does not give appropriate result about direct tensile strength. The strength is due to mixed stress field and orientation of fibres. The failure pattern of the concrete gives clear idea about the ductility of the material. Failure patterns of split tensile test indicate that a specimen after first cracking does not separate due to failure of the structure. The split tensile strength test results for conventional concrete and different percentage of polypropylene fibres as shown in table 4.6.

Sample	**Load in KN**	**Fibre Dosage**	**Split Tensile Strength in N/mm^2**

	7 Days	28 Days	in kg/m³	7 Days	28 Days
Conventional concrete	80	170	0.0	1.13	2.44
20% fibre replaced concrete	150	180	0.38	2.12	2.55
40% fibre replaced concrete	100	130	1.09	1.41	1.83
60% fibre replaced concrete	90	110	1.79	1.27	1.55

Table 4.6 Comparison of split tensile strength

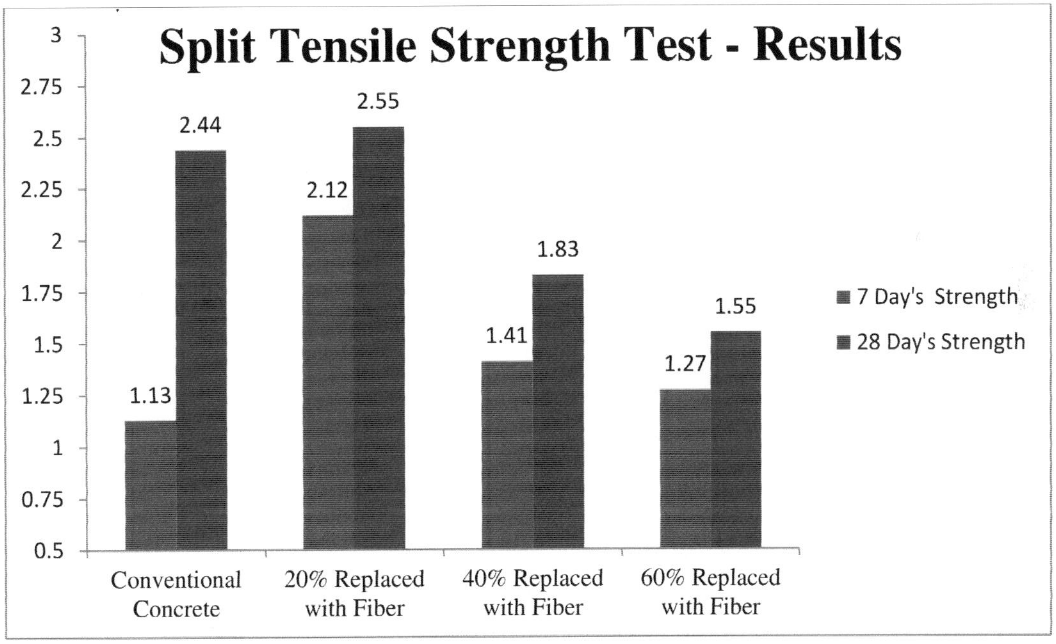

Fig 4.6 Comparison of split tensile strength

From the result, it has been observed that 7 days and 28 days split tensile strength of the concrete with the dosage of 20% fibre replaced concrete is higher than conventional concrete. Further dosage of fibres to the concrete mix reduces the tensile stress of concrete.

COMPARISION OF FLEXURAL STRENGTH: Nominal increase in strength was observed for all the proportions of fibres when compared to normal mixes. The improvement in flexural strength is recorded due to the improvement in mechanical bonding between the cement paste and fibre. As amount of fibre increases in mix, it extensively helped to reduce the widening of crack more effectively, hence resulted in increase in flexural strength. In order to execute further, to increase the crack opening the applied load need to be increased, this leads to the formation of another crack. The split tensile strength test results for conventional concrete and different percentage of polypropylene fibres as shown in table 4.7.

Sample	**Load in KN**		**Fibre Dosage in kg/m^3**	**Flexural Strength in N/mm^2**	
	7 Days	**28 Days**		**7 Days**	**28 Days**
Conventional concrete	5.5	6.0	0.0	2.75	3.0
20% fibre replaced concrete	5.9	6.5	0.35	2.95	3.25
40% fibre replaced concrete	5.0	4.5	1.01	2.5	2.25
60% fibre replaced concrete	4.0	3.5	1.67	2.0	1.75

Table 4.7 Comparison of flexural strength

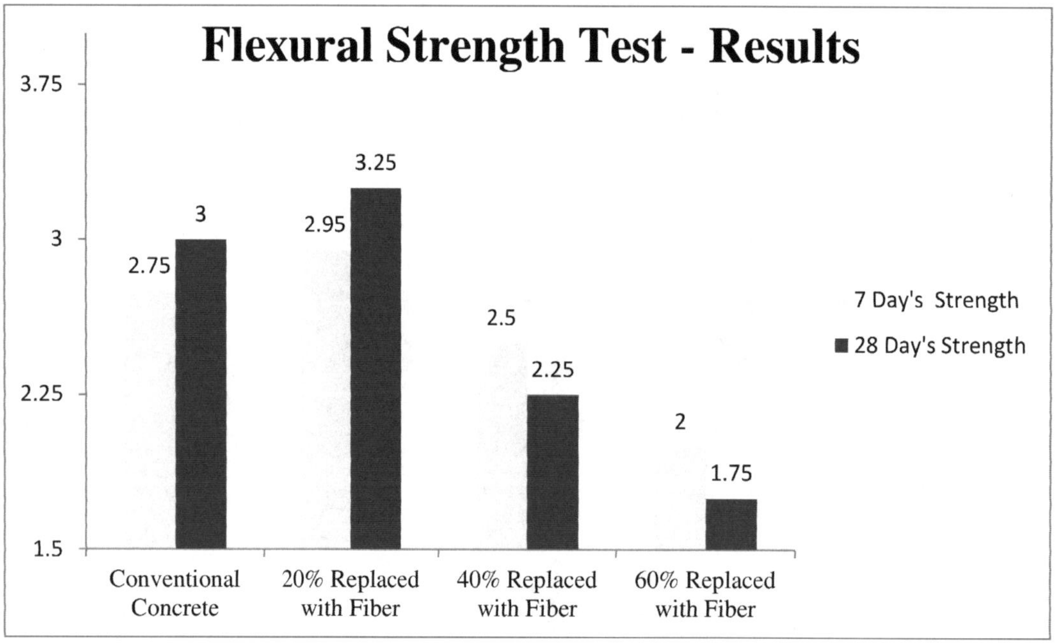

Fig 4.7 Comparison of flexural strength

It has been observed that the addition of polypropylene fibres with 20% dosage to the mix increases the 7 days and 28 days flexural strength of the concrete confinement provided by fibres. Further addition of fibres reduces the strength of the specimen.

5. CONCLUSION

The automobile plastic waste is a threat to the environment when it is disposed; therefore it is used effectively in the concrete mixture as fibre to improve the strength and workability of concrete. The fresh and hardened properties of conventional concrete are compared with the various percentage additions of fibres in concrete. From the test result the following observations are made,

- The workability of conventional and partially replaced polypropylene concrete are good and satisfies the limit
- The addition of polypropylene fibres of about 20% actually increases the 7 days and 28 days compressive strength. When the replacement is 20% the compressive strength is found to increase by 10% than conventional concrete. When the proportion of fibre is increased to 40% the strength decreases by 3% to 6% than conventional concrete.
- The tensile strength gets increased to 5%~10% when fine aggregate is replaced by 20% of fibres. Further addition of fibres gives lower strength than the conventional concrete.
- The flexure strength gets increased to 7%~8% when 20% fibres are replaced the fine aggregates in concrete. After which strength starts reducing with further addition of fibre in concrete. From the results we can conclude that the proportion of fibre added should be limited to maximum of 20%. If the higher percentage is added the negative effect is observed and the strength of concrete degrades. Thus the proper percentage composition of polypropylene fibre used with concrete will improve the strength and workability of concrete.

6. REFERENCE

[1] Ammar A.Muttar 2013 *Improving the mechanical properties of no fine aggregates* Volume 21 Journal of Babylon Univesity / Engineering science / No 2.

[2] T.Abadjieva P Sephiraj 2005 *Investigation properties of no-fine concrete, Property of concrete* pp 629 to 632.

[3] Anne Laning 1992 Synthetic fibres primarily used to reduce shrinkage cracking The Aberdeen Group Publication 1992.

[4] Faisal Fouad Wafa 1990 *Properties and application of fibre reinforced concrete* Volume 2 JKAU:Engineering science pp 49 to 63.

[5] Islam M.A Rahman M.M Ahmed M 2011 *Polymer modified concrete* The Indian concrete journal pp55 to 63.

[6] Kolli .Ramujee 2013 *Strength of polypropylene fibre concrete* Volume 2 Issue 8 International journal of Innovative research in science engineering and technology pp 3409 to 3413.

[7] Mehul J.Patel Kulkarni S.M 2013 *Effect of polypropylene fibre on the high strength concrete* volume 2 Issue 2 Journal of information , knowledge and research in civil engineering.

[8] Pakravan H R Jamshidi M Latifi M Pache F 2016 *Cementitious composites reinforced with polypropylene and polyacrylonitile fibre* US: Elsevier science publishing Ltd

[9] Saeed Ahmed, Imran A Bukhari, Javed iqbaul siddiqui, Shahzad Ali Queshi 2016 *A study properties of polypropylene fibre concrete* 31st conference on our world in concrete and structure, Singapore , CI-premier PTE Ltd.

[10] Sivaraja M Kandasamy S Thirumurugan A 2010 *Mechanical strength of fibrous concrete with waste rural materials* volume 69 Journal of scientific and industrial research.

[11] VanGemert D Czarnecki L Lukowski P Knapen E 2004 *Cement concrete and concrete polymer composites two merging world* A report from 11th ICPIC congress in Berlin.

[12] Youjiang wang Stanley backer Victor C L I 1987 *An experimental study of synthetic fibre reinforced cementitous composite* Journals of material science 22 Chapman and Hall Ltd 4281-4291.

[13] Bert-Okonkwor C.B.N Okolie K.C Ezeokoli F.O Ohazulume G.C 2020 *Comparative Analysis of Concrete Strength Made from Selected Brands of Cement in Anambra State, Nigeria* Archives of Current Research International.

[14] Merbouh, M'hammed, and Bachir Glaoui 2015 *Plastic Waste in Cement Concrete, LDPE and PVC Introduction Effect* Advanced Materials Research.

[15] Noor Hazlin Abdul Rahim, Noridah Mohamad, Abdul Aziz Abdul Samad, W. I. Goh, Norwati Jamaluddin 2017 *Flexural Behaviour of Precast Aerated Concrete Panel (PACP) with Added Fibrous Material: An Overview* MATEC Web of Conferences.

[16] Iwona Frydrych Elangovan G Mohan Das K 2018 *Flexural Behaviour of Fibre Reinforced Concrete Beams with Different Aspect Ratios* Fibres and Textiles in Eastern Europe.

[17] M.S. Shetty 2013 *Concrete technology theory and practises* S.Chand publications.

[18] IS 456:2000 2007 *Indian Standard, Plain and Reinforced Concrete – Code of Practice* Fourth Revision.

[19] IS10262: 2009 *Indian Standard, Concrete Mix Proportioning – Guidelines* First Revision.

PEG 400 EFFECT ON PROPERTIES OF SELF CURING CONCRETE

Sudharson.G[1], Kalpana M[2], Anburaja D[1] and Prathap M G[3]

1. ***Assistant Professor, Department of Civil Engineering, Saveetha Engineering College, Chennai, TamilNadu, India***
2. ***Associate Professor, Department of Civil Engineering, Saveetha Engineering College, Chennai, TamilNadu, India***
3. ***Research Scholar, Department of Civil Engineering, SRM Institute of Science and Technology, Kattankollathur campus, TamilNadu, India***

ABSTRACT- Curing becomes important to concrete to make it to achieve the required strength. In order to obtain strength at early stages we introduce the concept of self-curing concrete instead immersing or spraying on it. Water solvent polymers serve as self-curing agents for concrete. In the present study, PEG 400 is used to discover its effects on strength (Compressive, tensile and flexure) of concrete, adding it at different rates to the weight of cement up to 2%. The grade of concrete used here is M30. 1% of PEG 400 curing agent was found to be optimum in determining the strength properties. At higher rates there is a slight decrease in the strength value. The aim is to evaluate the performance of water-soluble polymeric glycol as curing agents. The advantages of curing agents serve its purpose in desert areas where water is not adequately available.
Key words: *Self curing concrete, curing agent, water solvent polymers, strength properties and PEG400*

INTRODUCTION

For more effective hydration of cement and reduced self-desiccation the need of moisture becomes important. This need will be fulfilled by providing proper self-curing or internal curing technique. The quality of concrete is affected by the loss of moisture content on exposure to the external environment. The factors that affect it are velocity, relative humidity, atmospheric temperature, water cement ratio and type of cement used. In the initial stage evaporation leads to plastic shrinkage and later leads to drying shrinkage. Continuous evaporation happens due to chemical potentials of water i.e. moisture content. Curing temperature is major factor contributing to strength development. The polymers added will reduce this chemical reaction and helps it maintain its moisture content. The usage of self-curing admixture becomes important at this point. This study will focus on comparing this with conventional concrete and finding the way to increase its strength. Scope is Investigate the effect of PEG 400 on strength and other properties of of concrete

OBJECTIVES

- Compare fresh and hardened properties of self curing concrete with conventional concrete with varying % of self-curing agent
- Investigate the mechanical properties of concrete i.e. compressive tensile and flexural strength in varying percentages of PEG 400.
- Find optimum dosage of shrinkage reducing admixture of PEG 400 for M30 concrete grade
- In this study the effect of two different curing agents has been examined to compare its performance for optimization.

IOP Conf. Series: Materials Science and Engineering 1026 (2021) 012014 doi:10.1088/1757-899X/1026/1/012014

METHODOLOGY

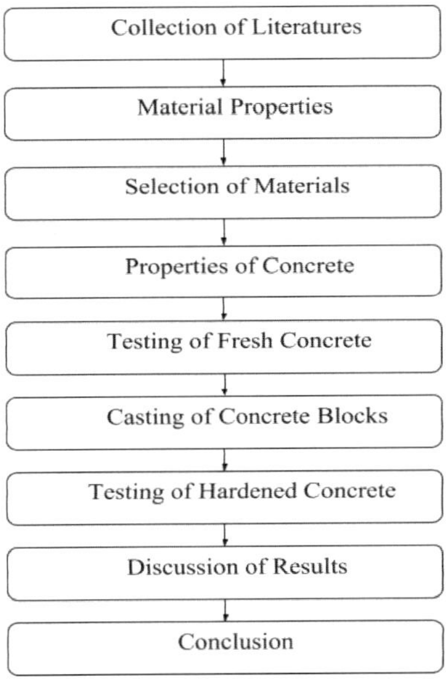

RESULTS AND DISCUSSIONS

COMPARATIVE TEST RESULTS OF CEMENT

SI No.	Test on Cement	Value Obtained	Permissible Value
1	Specific gravity	3.16	3-4
2	Fineness test	2%	10%
3	Soundness test	1 mm	Max 10mm
4	Standard consistency test	32%	-
5	Initial setting time	45 minutes	Min 30 minutes
6	Final setting time	540 minutes	Max 600 minutes

GRAIN SIZE DISTRIBUTION OF FINE AGGREGATE

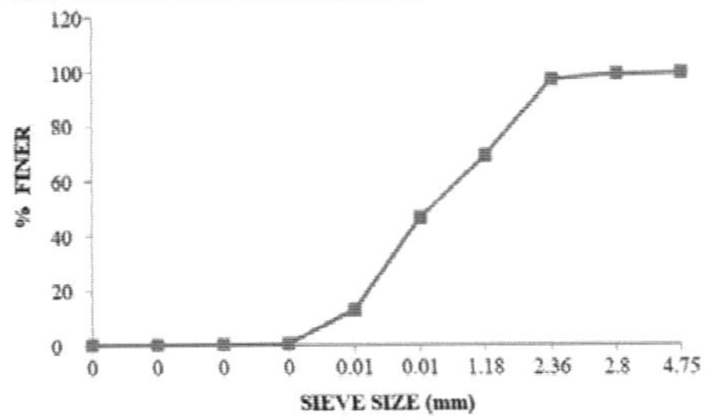

COMAPARTIVE TEST RESULTS OF COARSE AGGREGATE

SI. No	Test	Coarse Aggregate	Permissible Limit
1	Specific gravity	2.6	2-3
2	Water absorption	3.48%	≤20%

IOP Conf. Series: Materials Science and Engineering 1026 (2021) 012014 doi:10.1088/1757-899X/1026/1/012014

SPECIMEN LABELS

Table shows the various specimens casted and their labels for the reference.

Sl No.	Specimen Name	Specimen Label
1	Normal concrete	NM
2	Concrete mixed with 1% of PEG	SCC1
3	Concrete mixed with 1.5% of PEG	SCC1.5
4	Concrete mixed with 2% of PEG	SCC2

RESULTS OBTAINED ON COMPRESSIVE STRENGTH TEST

Compressive Strength Results of NC

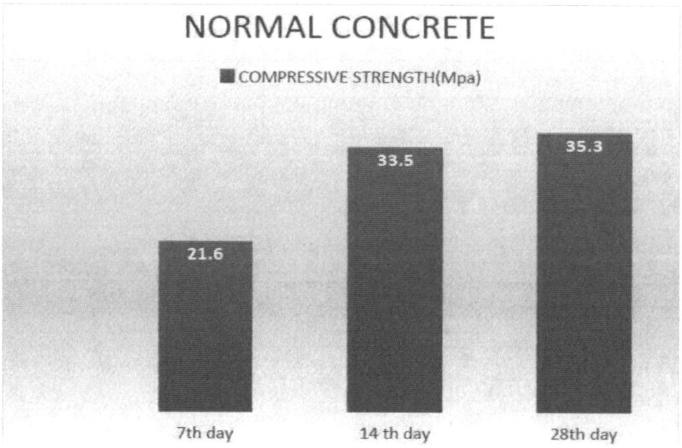

Compressive Strength test results of SCC of 1% replacement

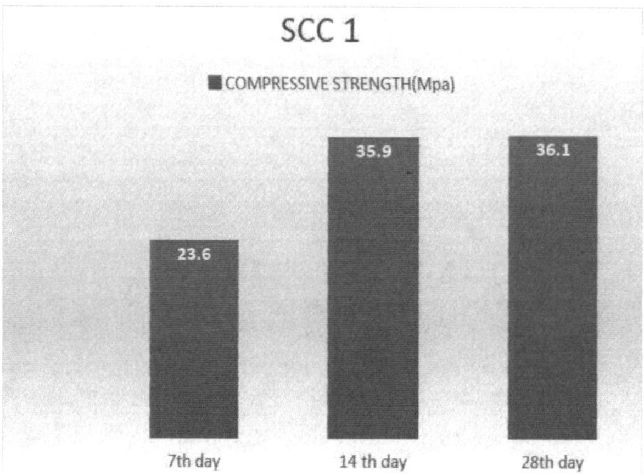

Compressive strength test results of SCC of 1.5% replacement

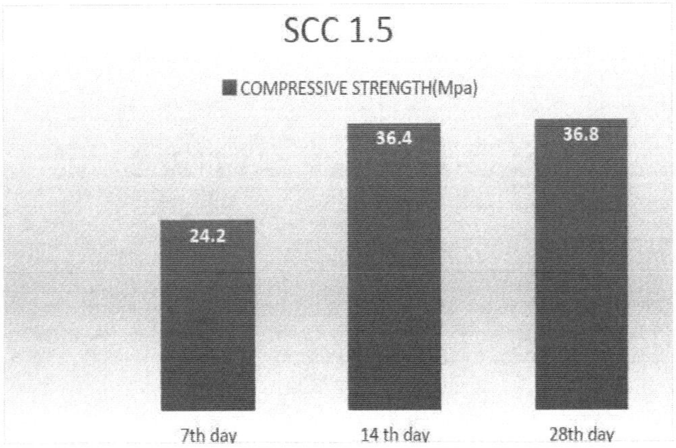

Compressive strength test results of SSC of 2% replacement

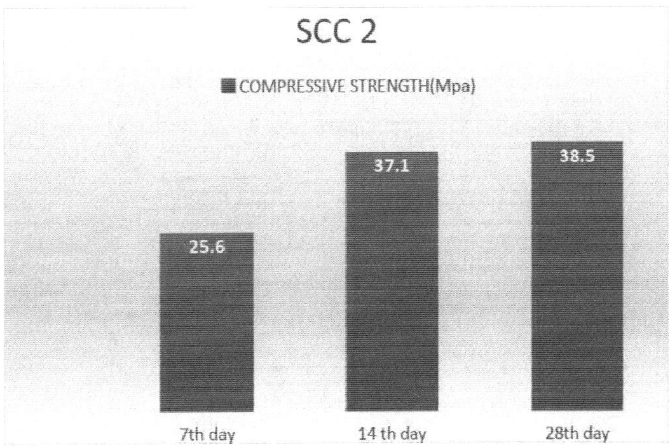

COMPARISON OF RESULTS IN CONCRETE

The results obtained from the compressive strength of concrete by the replacement of Polyethylene glycol. The compressive strength result for replacement of Polyethylene glycol with 1%1.5%, 2% was determined and compared with normal concrete.

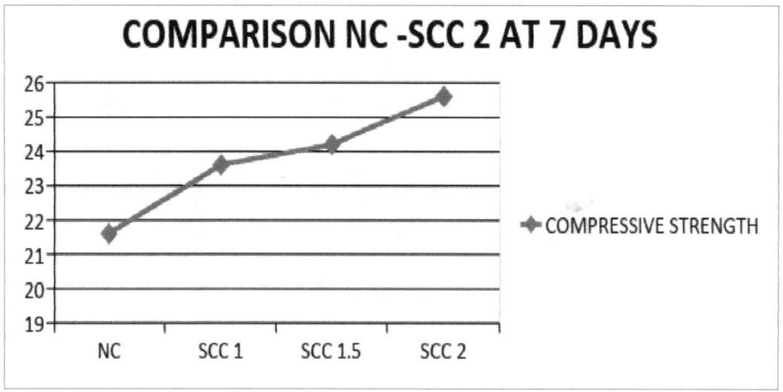

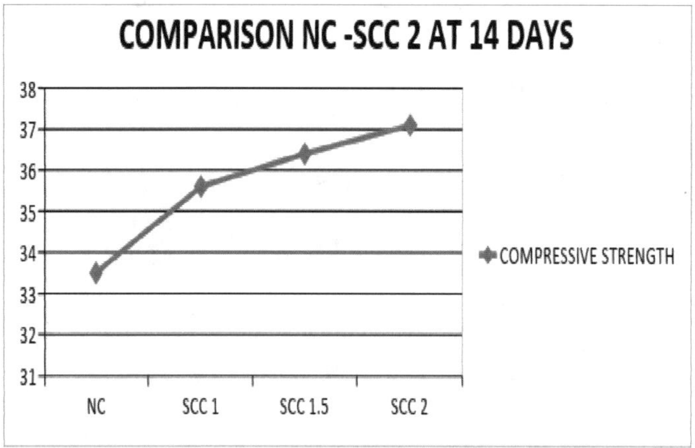

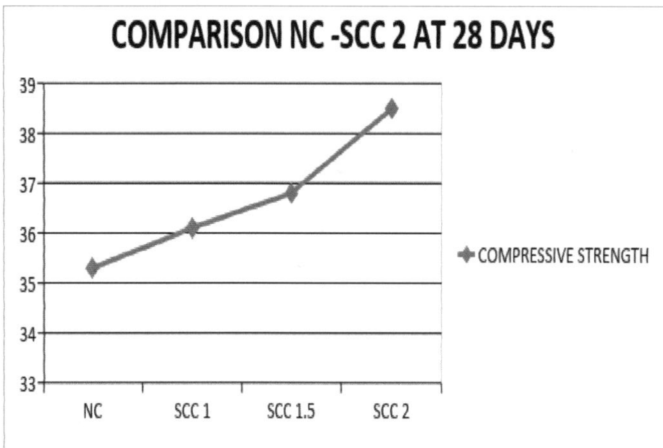

RESULTS FROM TENSILE STRENGTH OF CONCRETE

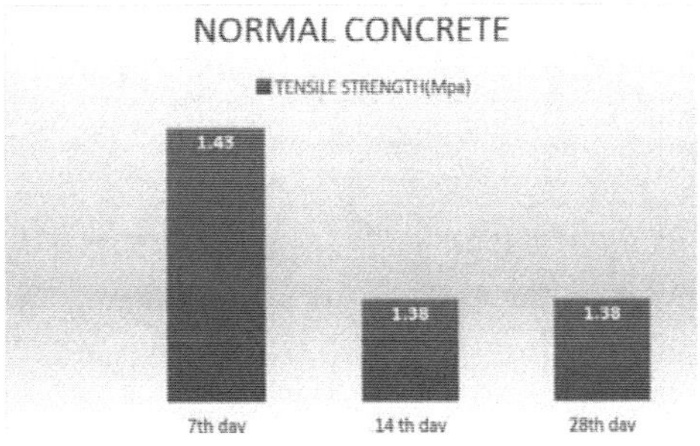

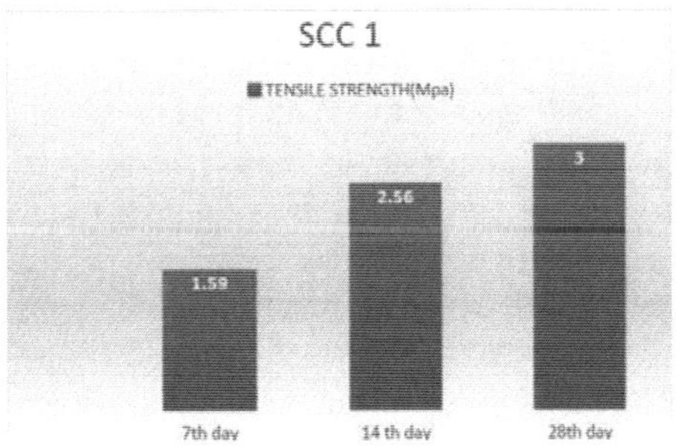

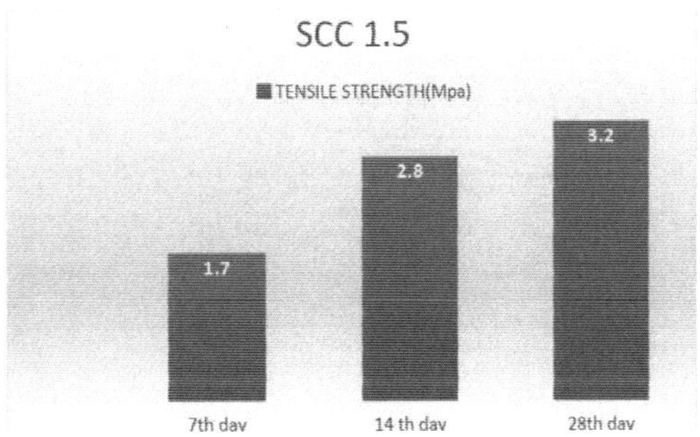

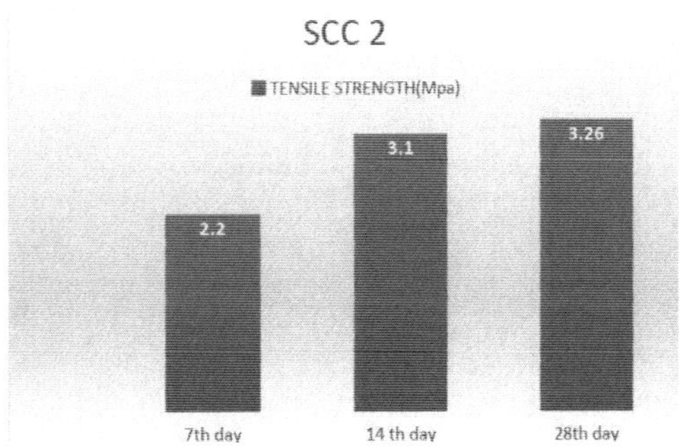

IOP Conf. Series: Materials Science and Engineering 1026 (2021) 012014 doi:10.1088/1757-899X/1026/1/012014

Comparison of results of SCC with NC at 28 days

Table 5.14 shows the results obtained from the tensile strength of 14 days test. Figure 4.2 shows the graph of SSC compared with NC for tensile strength of 14 days.

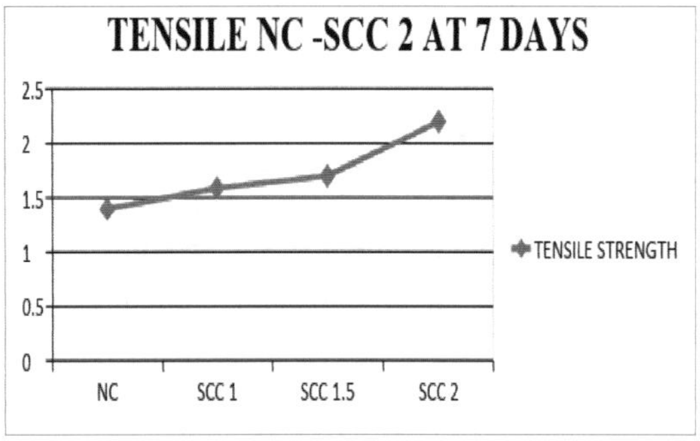

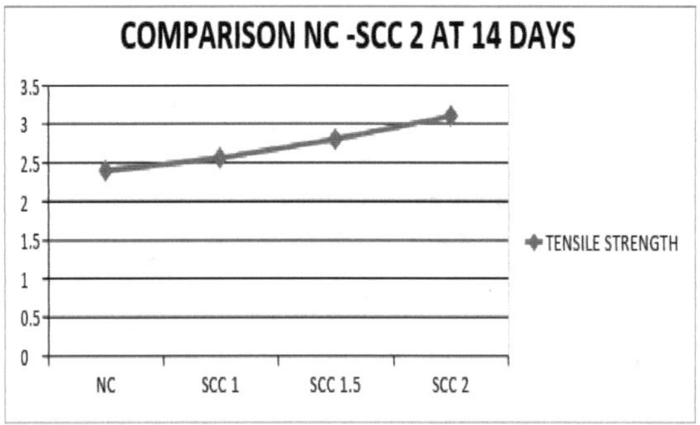

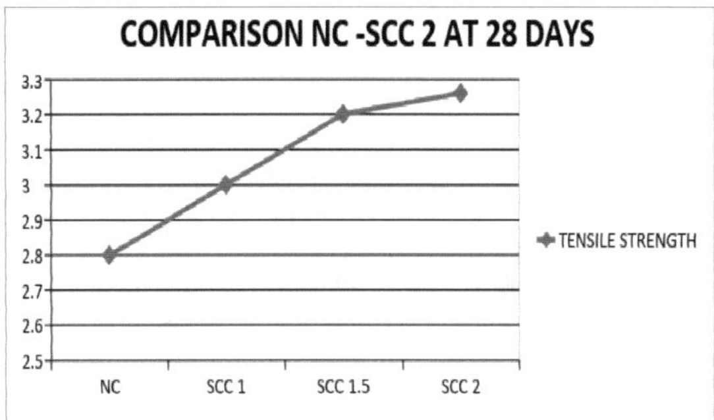

RESULTS OBTAINED ON FLEXURAL STRENGTH TEST

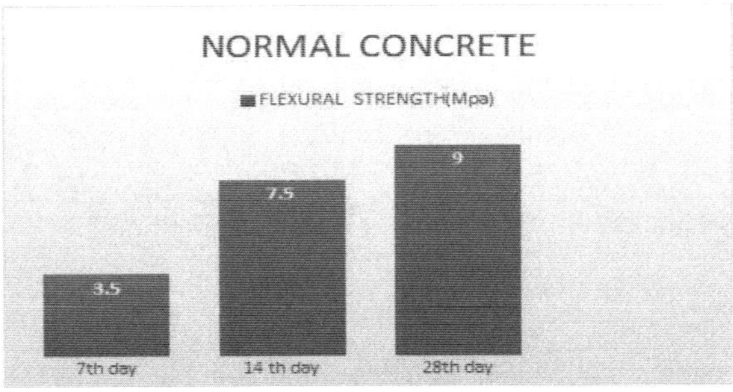

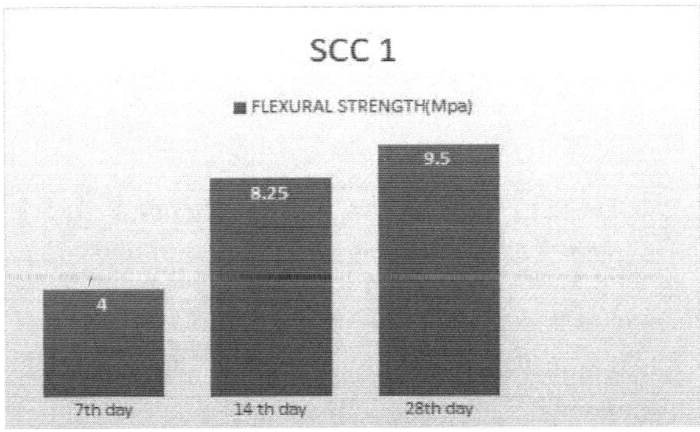

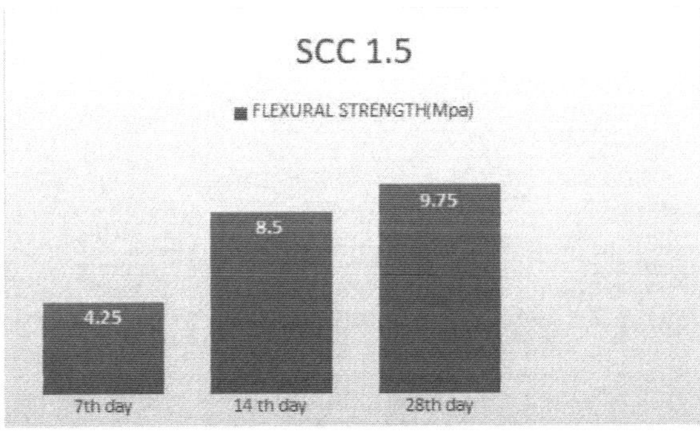

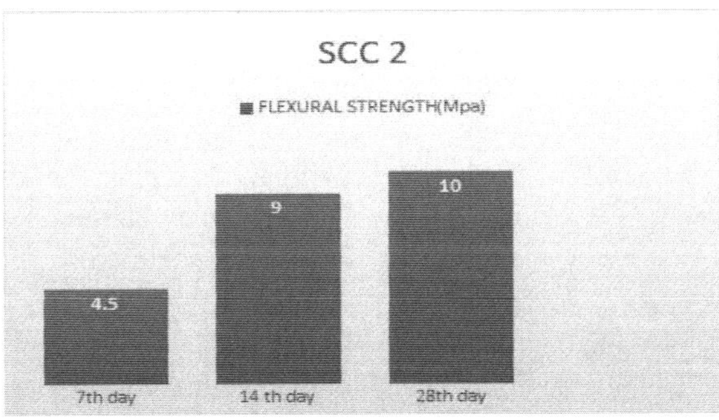

IOP Conf. Series: Materials Science and Engineering 1026 (2021) 012014 doi:10.1088/1757-899X/1026/1/012014

COMPARISON OF RESULTS IN CONCRETE

The results obtained from the compressive strength of concrete by the replacement of Polyethylene glycol. The Flexural strength result for replacement of Polyethylene glycol with 1%1.5%, 2% was determined and compared with normal concrete.

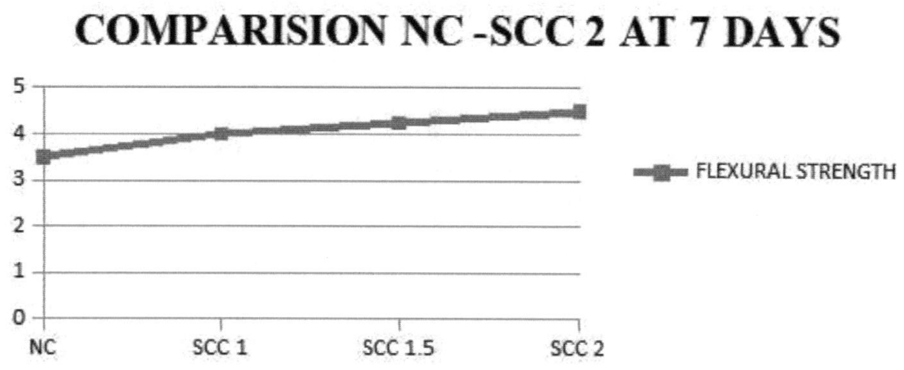

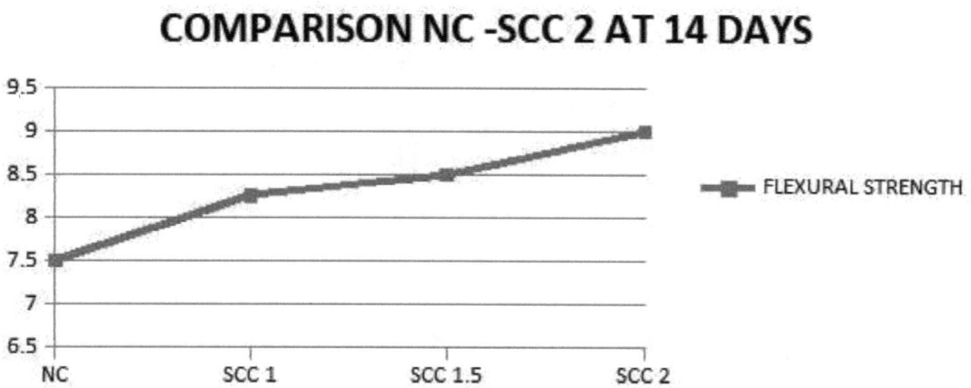

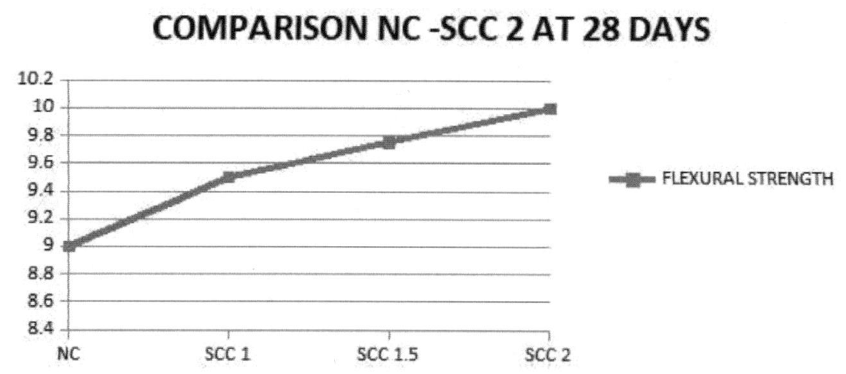

CONCLUSION

The results obtained, shows that PEG 400 concentration increases the slump value 0(**17.3 for scc 2 concrete**). Slump value increases as the concentration of PEG400 increases. Self-Curing concrete Compressive strength was found to be greater than conventional concrete. **The maximum strength values were obtained in scc 2 samples**. Compressive strength 38.5 Mpa and tensile strength is 10 Mpa. This pays a forward way to scc to have its application than conventional concrete in desert areas where we have less water for curing. M30 grade concrete with 2% PEG gives the same value of conventional concrete. This the optimised dosage of PEG value. Using this we can achieve good performance of concrete.

REFERENCES

1 Ahmad Mustafa Sabaoon, Navinderdeep singh, Jan-2018. *A Review Paper on Self Curing Concrete,* Volume: 05 Issue: 01 |
2 Akshara O.S, Divya Sasi, October-2016, *An Experimental Study on Mechanical properties of Self Curing Concrete,* Volume 7, Issue 10,.
3 Ananthi A, Ranjith R, Swarna Latha R, Vimal Raj R Jan 2017, *Experimental Study on the Properties of Self-Curing Concrete*, Volume 3: Issue 1,
4 Arunachalam Ananthi Jan.2017, *Experimental Study on the Properties of Self-Curing Concrete*, Volume 3,Issue 1.
5 Basil M Joseph Dec 29,2017 ,*Studies On Properties Of Self-Curing Concrete Using Poly-Ethylene Glycol*,
6 Hemanth Kumar. B,Radhakrishna Murthy. P February 2017, *A Brief Study on an Effect of Polyethylene Glycol (PEG-400) On Self Curing Concrete,* Vol. 6, Issue 2,
7 Gopala krishna sastry KVS, Puttur manoj kumar an 2019, *Self-curing concrete with different self-curing agents,* Volume- 330,
8 Magda,Mohamed,G. Mahdy,H. Abdel-Raheem,Z. Yehia 31 March 20150, *Self-curing concrete types; water retention and durability,* Issue 1,
9 Malindu Sandanayake, A. Yenduri, P.V. Indira and G. Zhang,Structural July 2016, *Performances of self-curing concrete using recycled coarse aggregate,* Vol. 90, Issue 00,.
10 Mohammed Junaid April 2015, *Self-Curing Concrete with Shrinkage Reducing Admixture*, Volume 2.
11 M.V.Jagannadha Kumar, M.Srikanth, K.Jagannadha Rao Sep-2012, *Strength Characteristics Of Self-curing Concrete*, Volume: 01 Issue: 01